WINE
SCIENCE

WINE SCIENCE

The Application of Science in Wine
from Vine to Glass

BY JAMIE GOODE

MITCHELL BEAZLEY

To LG

Third edition published in Great Britain in 2021 by
Mitchell Beazley, an imprint of

Octopus Publishing Group Ltd
Carmelite House
50 Victoria Embankment
London EC4Y 0DZ

www.octopusbooks.co.uk

An Hachette UK Company

www.hachette.co.uk

Picture acknowledgments
192 courtesy Della Toffola Pacific
All other photographs courtesy Jamie Goode

ISBN 978-1-78472-711-6

A CIP catalogue copy of this book is available from
the British Library.

Printed and bound in China

10 9 8 7 6 5 4 3 2

This edition
Group Publishing Director: Denise Bates
Art Director: Yasia Williams-Leedham
Design: Jeremy Tilston
Senior Editor: Alex Stetter
Indexer: Gillian Northcott Liles
Senior Production Manager: Katherine Hockley

Second edition
Commissioning Editor: Hilary Lumsden
Design: John Round Design
Senior Editor: Julie Sheppard
Editors: Jamie Ambrose, Constance Novis

Contents

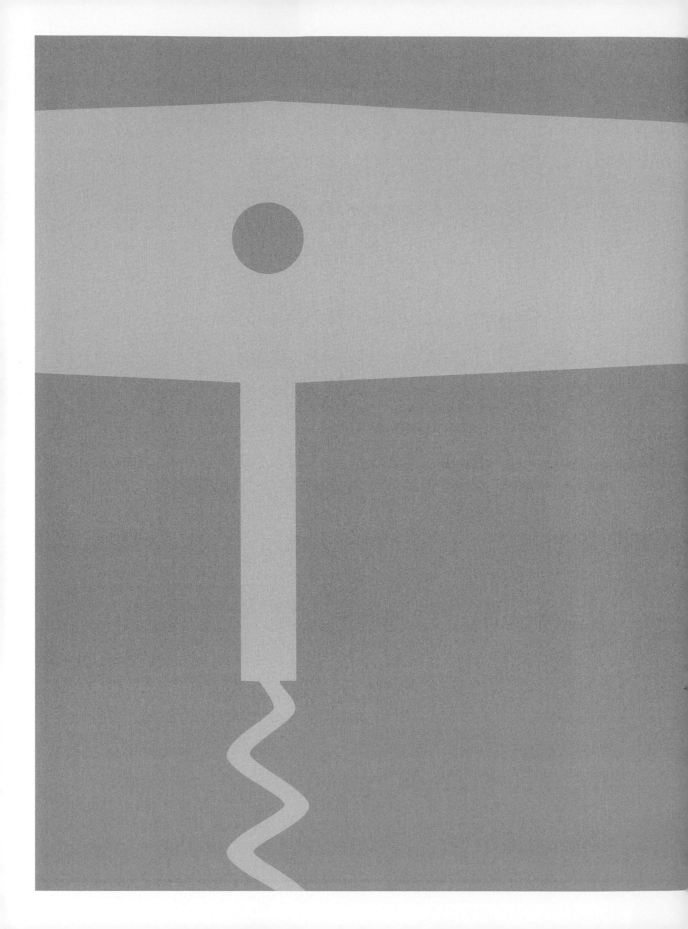

Note on the third edition

This is the third edition of *Wine Science*, which was first written in 2004 (published in 2005), and then revised in 2013 (published in 2014). In the seven years since the last revision, things have continued to evolve in the wine world, and I've tried to capture some of the most exciting of these changes here. I am happy and a little surprised by the success of this book. It seems that science interests a lot of people—even those who come from an arts background.

This is not meant to be a textbook, covering the whole of wine science in a methodical manner. Instead, I have set out to tell wine science stories in a way that would engage people who are not overly scientifically literate. I have also tried to make it current, addressing science-related issues that had not yet been written about widely for a broader audience. Most of all, I have tried to bring the perspective I have gained from extensive travels around the world of wine. As with the second edition of *Wine Science* I have added quite a bit of new material, but without expanding the book too much (the 100,000-word limit imposed by the publisher is a sensible one—there are too many overlong wine books out there). I also wanted to preserve the accessibility and readability of the first edition. I have added new chapters and taken some out. I have rewritten the remaining chapters, some quite extensively. Areas that have been covered in my recent books—*I Taste Red* (the perception of wine) and *Flawless* (wine faults)—have been trimmed down. There is increased coverage of winemaking processes, and the section on microbes has been expanded. Overall, around half the book is new.

First and foremost, I am a lover of wine. I have been well and truly bitten by the wine bug, and I hope that through the book, the fascination I have for this most remarkable of drinks is the dominant theme that places the scientific content in context.

Finally, a necessary note on conflicts of interest. Be assured that the opinions expressed in this book are my own genuinely held ones, and the topics covered here are those that I consider to be the most interesting and relevant. In weighing up the science I have reported, I have tried to be as evenhanded as possible. But I do need to mention some potential conflicts of interest, which readers deserve to know about. As well as writing about wine, over the last few years I have done some paid work for a range of wine-related businesses. This work includes running tastings, giving talks, taking part in brainstorming sessions, chairing debates, and acting as a discussion panelist. I am, however, not retained by any company. The companies I have worked for that are relevant here are Lallemand (yeast and oenological product company) and various generic bodies from different wine regions (Bourgogne, Centre Loire, New Zealand, Washington State, Portugal, Austria, Canada). It is important that readers know this so that they can make up their own minds.

Finally, since I wrote the first edition, social media has become part of the world we live in. Therefore I am very happy to interact with readers via twitter (@jamiegoode) and Instagram (@drjamiegoode), as well as good old-fashioned e-mail (jamie@wineanorak.com).

Jamie Goode, August 2020

Introduction: why wine science?

Wine is remarkable. Consider the following questions and statements. How can this drink of fermented grape juice have assumed such an important place at the center of many cultures, and maintained this place through millennia? How can it have spread from its origins in Eurasia some 8,000 years ago to become a frequent fixture on dinner tables across the world? People collect it, read books about it, spend large chunks of their disposable incomes on it, and some even give up their well-paid day jobs to go and make it. It has even survived (so far) the technological advances of the 20th century and the shift from largely rural-based economies to city living. Despite their best efforts, the branders and marketing wizards of modern retailing haven't been able to kill it. In non-wine producing countries it has begun to shed its predominantly elitist image and shows signs of becoming the drink of the masses.

From just one species of vine, *Vitis vinifera*, thousands of different varieties have emerged, each with their own characteristics.[1] The grapevine even has the capacity of transmitting some of the character of the site on which it is grown into the wine that it produces. As well as making drinks with myriad flavors, textures, and degrees of sweetness and astringency—many of which make perfect foils for different foods—the vine gives us a naturally alcoholic product with pleasant mind-altering and mood-mellowing characteristics.

While this book is about wine, its focus is to explore this remarkable substance through a particular lens—that of science. As an ancient drink, wine has been produced through the ages without the help of a modern scientific worldview. And many will argue that what science has brought to wine hasn't really helped it at all.

Some will go so far to suggest that the so-called "advances" promoted by scientists, such as the use of pesticides, herbicides, and mechanical harvesting to help in the vineyard; and filtration, cultured yeasts, enzymes, and reverse osmosis machines in the winery, have actually been detrimental to wine quality. Certainly, there is little doubt that the potential interventions that science has made possible have been abused.

But science has a lot to offer wine, at all levels, from industrial production of mega-brands to artisanal, handcrafted boutique wines. In this introduction, I'm going to outline why I think science is a fantastically useful tool for winegrowers. Like all tools, though, it can be used correctly or abused. Indeed, one of the goals of this book will be to show how it is possible to integrate many of the most interesting and absorbing topics

1 I am aware that this is biased toward *Vitis vinifera*, and that there are many fine grape varieties with some genetic input from American and Asian vine species. Hybrids can make good wines, and in the future the new resistant varieties that are being bred with some non-vinifera genes could play an important role.

How science works

The scientific community is a remarkable global enterprise. Researchers across the world are united by a common currency—data published in peer-reviewed scientific journals. It's an inclusive club, open to anyone, as long as they have good data and are prepared to play by the rules. How does it work? Scientists are employed by universities, government institutions, or private companies. The former will typically be paid a salary but will need to fund their work by means of grants, usually awarded by government-supported funding bodies or industry. To gain credibility and status, researchers need to publish their work in reputable peer-reviewed journals, and their publication record is how they are assessed. There are many thousands of these journals, and they vary in their scope from broad to very narrow. Not all journals are created equal, and some have much higher reputations than others. Typically, a scientist (or more commonly, a group of researchers) will write up their results and then choose the most appropriate journal to send them to. They will want to have them published in the highest-ranking journal possible (journals are ranked, for example, by the average number of times a paper published there is cited, with the status of the citing article borne in mind—it is called an "impact factor"), but they won't want to send their paper to a journal where it will be rejected, because of the delay in publication that will ensue. How do journals decide which papers to accept? This is where *peer review* kicks in, a process vital to the integrity of the scientific literature. Each journal has a board of editors made up of leading researchers in the field covered by the journal, and also a larger pool of scientists willing to act as referees for papers in their chosen subject areas. A paper coming in will be assessed by one of the editors: if it is clearly unsuitable it will instantly be rejected, but if it is potentially good enough, it will be sent out to two or more scientists for review. They will prepare a report on the paper, checking that it is correct, is suitable for the journal it has been submitted to (if it is a high-ranking journal, are the results exciting, novel, and significant?), and that the science is good. If they recommend it to be accepted, they might also suggest possible revisions or further experiments. Then the paper and the referees' reports are sent back to the editor, who makes a final decision whether to accept it, accept it with revision, or reject it. Getting your paper into one of the elite band of leading journals can make your career. It should be pointed out that peer review is a slightly controversial process because (1) it involves scientists reviewing the work of their peers who may well be their competitors, (2) it can take a long time, and (3) because some consider it not to be as rigorous as it should be since good papers are sometimes rejected while less good ones get through.

Science is highly competitive. The entry ticket into the scientific community is a doctorate (a PhD), which is awarded by a university for the successful completion of an acceptable thesis (a written account of research undertaken on the subject of choice). This typically takes from three to five years to achieve. But getting a PhD does not guarantee a research job. After you complete it you need to do what's known as a postdoc (postdoctoral research position), a short-term (usually three years) contract to work as a researcher in someone else's lab. After two such positions (preferably with one abroad), if you've been reasonably successful and have published several papers in good journals, then it's time to try to land a proper research job. These are few and far between, and competition for them is fierce. For those who succeed, though, running a successful research group is a highly rewarding career, albeit one that requires grueling hours and absolute commitment.

in wine with a scientific understanding of these issues, and that such an integration will assist in the production of more interesting, compelling wines at all levels. Even if your goal is to produce manipulation-free "natural" wine, a good grounding in wine science will help you reach this target with fewer disasters along the way. As an example, people pay a lot of money to buy wines coming from a particular patch of ground, or "terroir." Wine science will help us understand what is special about that vineyard site and may thus facilitate the identification of similarly endowed sites or help in the production of better wines from vineyards less blessed by nature.

SCIENCE IS USEFUL

The scientific method provides us with an incredibly useful toolkit. It helps us overcome our biases and prejudices and allows us to answer difficult questions. It helps us to be objective. It presents a coherent model of the world around us that assists our understanding of this environment and enables us to develop new technologies that actually work.

It needs to be emphasized that objectivity is one of the keys to the successful practice of science. By nature we are not objective. We are pulled and pushed in various directions by our built-in preconceptions, predilections, and prejudices, often subtly, sometimes not so subtly. We frequently display confirmation bias, recruiting pieces of evidence that fit with our narrative of the world around us. Good scientists will step aside and try, as much as is possible, to be ruthlessly objective about the phenomena they are studying. The two arms of scientific enquiry are observation and experiment. Scientists look at what's there, formulate hypotheses, and then test those hypotheses by experiment, trying their hardest to disprove them. This is the only way they can be sure that they are correct.

Let's make this practical. Imagine you had a novel chemical treatment that you thought would protect your vines from mildew. How would you test it? Well, you could try treating all your vines with it and then see how they do. There is a problem with this approach, though. If you get positive results, how do you know they are attributable to your treatment, and not, for example, to the benevolent conditions of this particular vintage? The answer is, you don't. This is where the scientific method helps.

A more rigorous and useful approach is to compare the treated population of vines with what scientists term a "control," in this case, a group of vines that have not been treated, or more precisely a group of vines that have been sprayed with an inert substance according to the same schedule as the test group, to rule out the possibility that it is the act of spraying that is having the effect, rather than anything specific to the chemical. So you split your vineyard into two and treat just one half. Still, there is a problem with this experiment. In any vineyard there will be natural variation, and any significant results might be because one part of the vineyard enjoys better conditions than the other (it might be slightly warmer or have different drainage properties. The answer? Subdivide the vineyard further into dozens of different plots, and randomize the treatment such that plots that are treated are interspersed with those that aren't in a way that evens out the environmental variation.

So do we go ahead? Not yet. Once we get our results we will need to know whether any beneficial effect is significant. That is, what is the likelihood that such a benefit could have been obtained by chance, through natural variation in the measured populations? This is where statistical analysis steps in. Statistics is intrinsic to any good experimental design. Good experiments should be designed from the start with statistical analysis in mind: how many replications (repeated observations) will be necessary to produce a significant result? This can be worked out in advance. Whenever you see a graph or table presenting experimental results, your first question should be: how significant are the differences between the control and experimental treatments?

The number of experimental replications needed depends on the variation in the populations being studied. The variation in a set of results is defined by a statistical term, standard deviation. It's not necessary to go into details about how this is calculated. All we need to know for our purposes here is that measures such as this allow scientists to work out whether their results are meaningful or not.

Let's take a slightly different example that will throw some more light on how scientists work and think. You suspect that drinking wine may be beneficial for health by protecting against heart disease, but how do you study this? For ethical and practical reasons it is rarely possible to do direct experiments. You can't easily isolate a group of people and vary just one parameter in their environment, such as whether or not they drink wine, especially when you are looking at a disease process that takes many years to develop.

You might want to start by doing animal experiments, looking at the cardioprotective effects of wine consumption on rats or rabbits kept under controlled conditions. The advantage here is that you can study the physiological effects of your treatment in depth; the disadvantage is that while animal models are helpful, mice, rats, and rabbits are different from people, a factor that significantly limits the utility of any knowledge obtained in this way. Another avenue of investigation might be to identify a specific physiological process involved in the development of human heart disease, and then study the effects of wine consumption on this "surrogate" process over a limited period in human volunteers, perhaps over a couple of days. Of course, identifying a reliable surrogate process or marker is the key here, and this is a nontrivial challenge.

Instead, you could study large populations over time and try to correlate behaviors, such as wine drinking with changes in health status,

such as the progression of heart disease. This is the science of epidemiology, and it was precisely such a study conducted in the 1950s by Sir Richard Doll that showed conclusively for the first time what many people had suspected: that smoking is harmful to health. The key issues here are recruiting large enough populations to produce statistically significant results, controlling for confounding (more on this in a moment), and having a relevant, easily measurable endpoint (for example, in the case of heart disease, whether or not a heart attack occurs). So, let's say you have decided to look at the influence of wine drinking on the incidence of heart disease in a population of 1,000 randomly selected adults, using the incidence of heart attack as your endpoint. You'd need to get the population to fill in a drinking questionnaire (and here is a source of potential error: most people will underreport the amount they drink), and then follow up the incidence of heart disease in the different groups (i.e. nondrinkers versus light drinkers versus heavy drinkers) over a period of time.

What if you find that wine drinkers have reduced levels of heart attacks? Then you'll need to show that the effect is a significant one by using statistics. But we're not finished there; it gets more complicated. Even if there is a significant association between wine drinking and the risk of heart attack, this doesn't prove that wine drinking protects against heart attacks. It might be that the population who choose to drink wine is associated with another trait that is linked to reduced risk of heart attack. For example, on average wine drinkers might also eat a more balanced diet, or have higher levels of gym membership, or smoke less. It's also well-known that low income correlates with poor health status, for a variety of unspecified reasons, and people on low incomes might be underrepresented among the population of wine drinkers. These effects are known as confounding, and they need to be controlled for. One way might be to balance the different study groups by socioeconomic status, or do a study solely within one profession, to iron out any major discrepancies. It's complicated, but unless you take these sorts of precautions you'll end up with an unreliable conclusion.

If you want to know about the health effects of wine, you might also try a clinical trial where the effects of wine are tested on a group of patients or healthy volunteers. The key to success is using a placebo treatment and blinding the study: not letting the subjects know whether they are receiving the actual treatment or the placebo. Variations on this theme include crossover trials, where groups are switched from the treatment to the placebo halfway through. Studies can also be separated according to whether they are prospective (looking at the effects of interventions over a period following the beginning of the trial) or retrospective (using already gathered data to look back in time from a known endpoint).

Then there's the issue of mechanism. Epidemiology can tell you that a certain intervention or environmental factor has a particular effect on a population, but then you'll want to know why. In the case of wine, if it is clear that moderate drinking protects against heart disease, then what is the biological mechanism? Is it the effect of alcohol, or the effect of another chemical component of the wine? To answer these types of questions scientists frequently turn to animal experiments, simply because doing the equivalent tests on people wouldn't be ethical. The goal is that by understanding the mechanism, drug development or other targeted medical treatments might be possible.

THE RISE OF ANTISCIENCE

But despite the evident utility of science, we live in a culture that is now marked by a strong antiscience sentiment. Back in the 1960s and 1970s scientists were largely revered. Now they are treated with suspicion. Part of the public disenchantment with science lies in the fact that people feel let down: science promised too much and couldn't deliver. The application of science has led to breathtaking technological advances that show no signs of losing pace. Moore's law—the idea that computer processing power doubles every couple of years—is still holding very nicely. When I wrote the first version of this book, in 2004, my cellphone could make calls and receive texts. As I write now, in 2020, I have a shiny smartphone that is a powerful computer and very able camera. My digital camera of 2004 is now looking very outdated when compared with my current Micro Four Thirds mirrorless camera.

But despite this, scientific progress hasn't led to the nirvana of a happy, disease-, and crime-free society. Medical advances against the chief killers in the west—cancer and heart disease—are slow and have included a large number of false dawns. Malaria is still the world's largest killer and our treatments have advanced little. Bacteria are increasingly resistant in the face of our armamentarium of antibiotics, to the extent that we are facing a very real crisis where people are dying from infections that a decade ago would have been easily treatable. Bringing a new drug to market is hideously expensive, with myriad legislative hurdles, and the pipeline of new drugs in development is looking a little short. Consumers, disenchanted by the medical profession's perceived limitations, have turned increasingly to largely unproven alternative therapies. Even where science offers solutions for problems of the present and the future, such as genetically modified (GM) crop plants, consumers are not sure they want them.

Perhaps we have expected too much of science, or maybe scientists themselves have been guilty of promising what they can't deliver. Science is a tool, and an incredibly useful one, but it is no more than that. Science can't address issues that belong in the realm of ethics, morality, religion, politics, or law. That scientists have sought to impose their ideas in these realms is not the fault of the scientific method, nor does it mean that science as a tool or process has failed. Instead, society has been wrong to look to scientists to provide enlightenment where it simply cannot. To use a rather far-fetched analogy, if we are going on a journey, science is the engine that helps get us there, but it shouldn't be driving the car. Scientists have often been guilty of undervaluing or ignoring things that cannot be measured. Let us be philosophical for a moment. Metaphorically speaking, many people would say that wine has a "soul." It is common to find people involved in the production of wine who have a strong sense that there is a "spiritual" element to what they are doing. They believe that they need to operate with integrity and produce honest wines that reflect a faithful expression of the sites they are working with.

Scientists typically find this sort of attitude hard to understand, because ideas like this can't be framed in scientific language. But isn't it best if we can establish some sort of dialogue between scientifically literate wine people and those who choose to describe their activities in other terms, such as proponents of biodynamics?

How does all this relate to wine? In this book I am going to be looking at wine through the particular lens of science. I'll be exploring how science is a useful (even vital) tool in the fields of viticulture, winemaking, and also in terms of helping us understand the human interaction with wine. But I am not suggesting for one minute that wine—this engrossing, culturally rich, life-enhancing, and enjoyable liquid—should be stripped of everything that makes it interesting and turned into an industrially produced, technically perfect, manufactured beverage. Science is a tool that can help wine, but this doesn't mean that wine should belong to the scientists. For this reason, I'll be leaving the familiar, safe ground that you might expect a book titled "wine science" to cover, and venture into, some of the more absorbing issues that get wine lovers talking.

COMPLETING THE LOOP

The goal of this book is to complete the loop. Let me explain. Wine is the result of a number of different applied scientific disciplines. If we are to understand wine science, we need to combine these disciplines, and it's this combination that I'm calling "completing the loop." First, we have viticulture. We need to understand the biology of the grapevine, and specifically how we can encourage it to produce grapes that can be used to make great wine. But more than this, we also need to understand the agroecosystem of the vineyard.

The old-fashioned view that simply saw the vine as the sole focus—emphasizing the crop plant but then treating the soil as merely a growth medium for the vine—is now out of date. The more we learn about the importance of the soil and its microlife, the more interesting the story becomes, so what we are seeking is a complete picture of how the vineyard environment interacts with the vine to produce wine grapes that have desirable properties for making wine.

It doesn't make sense, though, to consider viticulture in the absence of an understanding of the next the next stage: winemaking. Viticultural decisions are taken with consideration of the winemaking that is to follow. Questions such as yield, fruit zone leaf removal, pruning, trellising, and picking decisions all have an impact on wine style and quality. Good winemakers don't just receive fruit from their viticulturists. Instead, they are aware of what is going on in the vineyard and will likely be helping in making decisions. So, we join together viticulture and winemaking.

Now we enter the winery. The grapes are received, and winemaking begins. There are many choices to be made and the first significant one will have been the picking decision. The key interface between viticulture and winemaking is when the grapes are picked. An analysis of the grapes coming in will then influence subsequent choices, and the winemaker will have options, including whether or not to add sulfur dioxide (and, if so, how much) and then whether other additions are also needed. There are decisions with pressing or maceration, type of fermenter, fermentation temperature, whether or not to use cultured yeasts, and then the sort of vessels to use for fermentation and aging. There is a lot of science here, but also a fair bit of intuition and educated guesswork. Wine really is a blend of art and science.

Just as viticulture and winemaking merge, so does winemaking and the next part of the loop: sensory. This relates to how we, as humans, perceive wine. Throughout the winemaking process (and, usually, beginning in the vineyard), winegrowers will be using their sense of taste to assess the grapes, and then the must. They will also taste the various stages of fermentation.

The science of perception comes into play, and it is a very interesting subject indeed. We are not measuring devices, but we are very good at taste and smell. And if we are to make sense of perception, then we must understand how flavor is created in our brains. Flavor perception is multimodal, combining different sensory inputs in such a way that we are not aware of all the processing stages.

Sensory is where the loop is completed. After winemaking is finished, the wine is bottled and eventually makes its way to the consumer. We have another overlap here, since winemaking choices, and also viticultural decisions, are made with the end consumer in mind. Or, in an ideal world they should be. So sensory overlaps with viticulture and winemaking. There is also the science of understanding the consumer, and marketing.

Ultimately, as the wine is drunk, the loop is closed. And the goal of this book is to present you, the reader, with a detailed exploration of some of the most interesting topics in the world of wine science, presented in such a way that this loop is neatly closed. Science has a lot to offer wine. My goal is that in writing this book, which is designed to be accessible to nonscientists yet still with enough meat to keep the scientists engrossed, I'll have helped some to an enhanced understanding of wine that will assist them in their pursuit of this culturally rich and fascinating beverage.

Section 1
In the Vineyard

1 The biology of the grapevine

Agiorgitiko and Albariño, Baga and Bourboulenc, Cabernet Sauvignon and Chardonnay, Dolcetto and Durif: there are many thousands of different grape varieties, capable of making a bewildering array of different wines, but they are all cultivars of just one species: *Vitis vinifera*. Estimates are that across the globe there are some 14,000–24,000 different cultivars (the scientific term for variety), but because many of these are synonyms, these represent perhaps 5,000–8,000 varieties. An influential book on the subject, *Wine Grapes* (Robinson J, Harding J, Vouillamoz J; Allen Lane, London 2012), identified 1,368 varieties used commercially to make wine. This single species is the source of almost all the wine consumed today. *Vitis vinifera* is commonly referred to as the Eurasian grapevine, because of its origin in the Near East, at the meeting point of Europe and Asia. This is one of the places where *Vitis vinifera* can still be found growing wild today, as *Vitis vinifera* subspecies *sylvestris*.

The genus *Vitis* actually contains around 70 different species, many of which are found growing in the USA, such as *Vitis labrusca*, *V. riparia*, and *V. berlandieri*, and also Asia, which has its own set of native vine species. The native American vines are sometimes used to make wine and have been crossed with *V. vinifera* to make hybrids that are widely grown in places where disease pressure or climate rules out the vinifera varieties. And they have an important role to play in the breeding of new disease-resistant varieties (see chapter 7). But for now their chief significance lies in that many of them have evolved in conjunction with the aphid phylloxera, and so can coexist with it. As a result, American vines are used for rootstock onto which almost all vinifera vines are grafted. Without this, viticulture in Europe and much of the rest of the world would have been finished by the phylloxera epidemic that occurred in the late 19th century.

VINES IN THE WILD

When most people think about grapevines, they envisage pretty vineyards, with neat rows of vines arranged on a trellis system or grown as bushes. But this isn't how grapevines grow in the wild. Their natural growth form is as woodland climbers, using trees for support. Where the vine breaks through the canopy into sunlight, it flowers and produces grapes. These are eaten by birds, which then disseminate the seeds. Because of this growth form, vines need root systems that can compete for water and nutrients with the already established plants they are hitching a

Above A wild vine growing up a tree. This is the natural habitat for grapevines, which belong to a group of plants called lianas. Structural parasites, they are adapted for this function through the possession of tendrils.

ride on; the ability to make the most of limited resources is a prerequisite to this sort of lifestyle. Vines also need shoots that are capable of rapid elongation to grow toward the outside of the host canopy to find sunlight. Then, when they are in the light, this is the right time for the shoots to produce flowers and thus fruit, an effort wasted if it takes place in the shade of the canopy. This makes it clear that the vine is designed to be a highly competitive plant with a flexible growth form—vines have to adjust to the shape of whatever host plant they are growing on. Knowing what the vine is programed to do can help in uncovering the scientific basis of effective viticulture.

DOMESTICATING WILD GRAPEVINES

Ancient humans living in the right places no doubt would have been familiar with the wild grapevine and its attractive fruit. Mystery surrounds how the grapevine was first domesticated. One speculation, known as the Paleolithic hypothesis, seems plausible. Imagine some early humans foraging for food. They discover some brightly colored berries growing on vines suspended from the trees, so they pick them and eat them. They taste good, so these foragers collect as many as they can in whatever container they have on hand. On the journey home, the weight of the mass of grapes crushes a few, which then start fermenting. The result is a rather rough-and-ready wine that collects at the bottom of the container after a few days. If you found this sort of liquid mass at the bottom of a pot, you'd give it a try, wouldn't you? It's hard to imagine any wine produced in this fashion tasting wonderful, but then these folks probably weren't all that fussy. When they experienced to a small degree the mind-altering effects of this liquid, you can imagine it catching on fairly quickly. Deliberate planting of grapevines would have likely followed. Someone would likely have planted a few of the seeds, and with a little trial and error, have worked out how to make a vineyard. They would have selected for the rare wild vines that

had both male and female flowers (dioecious), because these would have been more fertile: in their native state, the wild vines would have been either male or female (monoecious). It is hard to be precise, but it is estimated that this grapevine "domestication" first occurred at least 7,000 years ago, and possibly as long as 10,000 years ago.

VINE STRUCTURE AND DEVELOPMENT

There are six main challenges facing plants growing out of water. First, find enough water, and then hang onto it, while at the same time being able to exchange gases with the atmosphere. Second, defend against being eaten by herbivores or destroyed by pathogens. Third, find enough light for photosynthesis. Fourth, reproduce and disperse. Fifth, adapt to seasonal rhythms and the variability in the environment. Sixth, deal with competition. It is the way that plants have met these challenges that has shaped and constrained their growth form and physiology. As well as these, the vine has further constraints and specializations resulting from its lifestyle as a structural parasite.

Let's take a look at how the grapevine works, beginning at the bottom with the roots. Roots serve two functions: anchorage and uptake. What exactly do vine roots take up from the soil? Like other plants, vines don't need much—they make everything themselves. But they do need an adequate supply of water and dissolved mineral ions, termed as macro- and micronutrients. These are inorganic (they don't contain any carbon). Root growth is determined by interplay between the developmental program of the plant and the distribution of mineral nutrients in the soil. The roots seek out the water and nutrients in the soil, sensing where they are and then preferentially sending out lateral roots into these areas. Low nutrient levels in the upper layers of the soil cause the roots to grow down deeper. This is likely to improve the regularity of water supply to the vine, and such roots can reach depths of nine feet or more. The root system of one vine is capable of supporting an enormous mass of aerial plant structure.

Above ground, the vine has a growth form well suited to life as a climber. Its shoot system is simple, adaptable, and capable of fast growth. The vine never intends supporting itself, so it does not waste resources on developing girth. Thin, long shoots are the order, which in turn can produce lateral shoots that eventually become woody. The formation of woody tissue is not for structural reasons but to provide protection, particularly during the dormant period. At regular intervals buds are formed. These buds are complicated structures, containing the potential for leaves, flowers, and tendrils, and develop over two seasons, with a rest over the dormant period.

SHOOT MORPHOLOGY

The stem is separated into sections by structures known as nodes. At each node a leaf is formed on one side, with either a tendril or a flower bud on the other. Thus both vegetative (leaf) and reproductive (flower) meristems (the growth region where cells are actively dividing) are formed simultaneously on the same shoot.

Above Bud burst, Nyetimber vineyard, Sussex, England. The start of the growing season is a nervous time for many winegrowers because young buds are vulnerable to spring frosts. Global warming has led to earlier bud burst in many regions, increasing risk.

Light is the key to vine growth. In the absence of light, shoots show negative gravitropism (meaning they grow away from the ground). But light is the overriding growth cue: shoots are positively phototropic, growing toward light. Light is also the chief cue for flowering induction. The tendrils are important structures for the vine's climbing habit. They are modified stems that coil around supporting structures.

BUD DEVELOPMENT AND FLOWERING

The flowering process in a grapevine is unusual, because it extends for two consecutive growing seasons. Flowering is first induced in latent buds during the summer, but initiation and floral development occurs the following spring. On flowering induction the shoot apical meristem (SAM; a meristem is a region in a plant where cells are actively dividing, producing new cells that have not yet developed into a specific cell type, known as undifferentiated cells) produces lateral meristems: these will either give rise to flowers or tendrils.

In grapevines, buds are formed and are first detectable in early spring in the axils (the inside of the join between the stem and leaf petiole) of the current year's leaves. These buds consist of several SAMs protected by bracts, which are scalelike in structure. The earliest-formed SAM usually develops as a lateral shoot while the rest remain dormant. In first months of development the SAM produces leaf primordia (cellular structures that will later give rise to leaves). Then, around May/June in the Northern Hemisphere, it produces lateral meristems (as opposed to apical meristems, meristems not at the tip or apex, but further down the plant) opposite the leaf primordia. The first two or three of these will produce flowers, the rest tendrils.

More meristems are produced such that by the end of the growing season the bud contains a shoot which has inflorescence (flower structure) meristems, tendrils, and leaves. These are all protected by bracts, scales, and hairs and become dormant in the fall.

During the following spring the bud is reactivated, and more meristems are produced. Crucially, the lateral meristems giving rise to inflorescences or tendrils are indistinguishable at the time they form. This keeps the options of the vine wide open, allowing environmental cues to dictate at this late stage whether these lateral meristems should become flowers or tendrils. The decision affecting this choice between tendrils and inflorescences has been the subject of a separate series of studies by researchers in Australia led by Mark Thomas.

Thomas and colleagues have shown that tendrils and inflorescences are structures that are similar in origin, and which can be converted either way by plant hormone application. Indeed, intermediate structures are commonly observed on vines. Their work has shown that gibberellins (a type of plant hormone) are major inhibitors of grapevine floral induction. In the natural habitat of the grapevine, a woodland climber, gibberellin would have two roles. It would promote the elongation of the stem and the production of tendrils, while at the same time suppressing fruit production. This suggests that there may

well be a connection between light sensing and gibberellin production (or responsiveness to gibberellin), because this is the sort of strategy that vines would use if they were in the shade of the host plant canopy. A vine in the shade would want to delay fruit production and maximize its energy on growing upward as fast as possible until it breaks through into the light.

When flowers are formed, their development and pollination occurs best during a period of warm, settled weather. Domesticated vines are hermaphrodite (meaning, each flower of each individual has both male and female structures) and can self-pollinate. Poor weather during this process can result in reduced or uneven fruit set, so flowering is one of the critical phases in the vineyard calendar.

GRAPE DEVELOPMENT

Grapes are for the birds. The wild vine "designed" grapes with the aim of enlisting the assistance of birds in spreading their seed. The kinds of places birds go are promising locations for dispersing seeds to. All those sugars in ripe grapes are a reward. But the grapevine doesn't want the birds to be carrying the grapes off too soon, before the seeds are ready for dispersal, or before the onset of fall, with its rains and favorable conditions for seed germination.

The maturation of grapes is therefore cleverly timed and separated into three phases. The first phase involves the development of the grape structure; throughout this process the grapes accumulate acids and experience rapid cell division. This growth then slows for the second phase, *veraison*, when red grapes change color from green to red and the skins of white grapes change from a hard green to a soft translucent green. The third phase follows this, when cell growth begins again and grapes accumulate sugar and phenolic compounds, and acids decrease. It's all for the benefit of the birds: unripe grapes are camouflaged (green color) and unappetizing

Above Flowering in progress, Columbia Valley, Washington State. This is a vital time in the vine's annual cycle. Bad weather at this stage can result in poor fruit set, and thus low yields.

Above Tendrils growing in Maipo, Chile. These are specialized structures that help the vine attach itself to the trees or bushes that it is growing up. The latent bud is capable of forming either flower clusters or tendrils. The two structures are related, and their developmental destiny depends on the weather (especially the quantity of sunlight) in the previous season.

(highly acidic, containing high levels of leafy-tasting pyrazines, and with harsh tannins), while ripe grapes stand out with their attractive red or golden color and taste lovely (the sweetest of all fruits, with lower acidity, riper tannins, and the herbaceous pyrazine flavors degraded or diluted). The grapes are telling the birds that dinner is served. Seeds are the first part of the grape to reach physiological maturity, around *veraison*.

GRAPE MATURITY

What is grape maturity in winemaking terms? It depends on the objective. Two types of maturity are talked about in wine circles: sugar accumulation and phenolic (or physiological) maturity, the latter also being referred to as flavor maturity by some. Match the right grape variety to your vineyard climate, and do your viticulture well, and you'll reach the goal of perfect maturation: flavor maturity coinciding with a sugar level that will yield a wine of some 12–13% alcohol. There is, however, a disconnect between the physiological processes that govern the rate of sugar accumulation and loss of malic acid,

which is dependent on climatic factors, and the color, aroma, and tannin development (phenolic maturity), which is nearly independent of climate. The result is that in warmer climates, grapes only reach physiological maturity at sugar levels that are considerably higher than in cooler regions.

Typically, in the cooler, classic Old World regions, grape harvest coincides with a shortening of day length and temperature, and thus sugar accumulation is more gradual. In these regions the measurement of sugar levels works well as a guide for when to harvest. It is also a simple measurement to make in the field. In these conditions, it is likely that by the time the grapes have accumulated 12 degrees of potential alcohol or so, they will have achieved satisfactory phenolic maturity. Pick at the same sugar levels in many New World regions and you'll end up with unripe flavors in your wines.

Light is crucial in the ripening process; grapes that are shaded contain less sugar and are more acidic than those exposed to sunlight. Light also affects bud fertility, so one of the key viticultural goals is therefore to encourage the vine to produce an open canopy, without dense, vigorous growth that could produce shading.

DIFFERENT VARIETIES, DIFFERENT CLONES

In taxonomic terms, *Vitis* is what is known as a genus, the taxon above species. Way back in evolutionary time, this genus split into three lineages. The Eurasian vine *Vitis vinifera* is a single species and is responsible for almost all wine made today. In contrast, there are dozens of different American species of *Vitis*, whose main importance in modern viticulture is to provide phylloxera-resistant rootstocks for grafting *Vitis vinifera* onto. Finally, there are a few Asian species of *Vitis*, of little importance for wine, but potentially an important source of resistance genes for breeding programs. *Vitis vinifera* may be just one species, but through evolution and numerous cross-fertilizations it has produced the thousands of different varieties that are used in winegrowing today. The effect of domestication has been largely to improve the fruitfulness of these various varieties and to distribute them widely.

So, what exactly is a grape variety? I posed the question to researcher José Vouillamoz. "A grape variety is the result of a seed that has grown into a plant that has been selected by humans, propagated over centuries or millennia by layering or cuttings, accumulating mutations along the way," he responds. "The older the grape variety is, the more diverse it is in terms of shape, color, qualities, and characteristics. All of these differences are what we call clones, and all the clones together make up the grape variety. The starting point is one father and one mother." Vouillamoz points out that a new variety cannot be made by a series of accumulated mutations. Instead, sexual reproduction is necessary. "It is just like a human being: every grape variety has a father and a mother. You can't say that by accumulating mutations you develop a new variety. Otherwise, where do you put the limit?"

By this definition, Pinot Noir, Pinot Gris, and Pinot Blanc aren't separate varieties. "Pinot is one single grape variety that is very old and has had many 'accidents' in its life—many mutations," says Vouillamoz. "Some of them were spectacular because they touched the color of the grape varieties and almost nothing else. When I do the DNA profiling of Pinot Noir, Pinot Gris, and Pinot Blanc, since I look at say ten or twelve different DNA regions, I don't see where the color mutation has happened. They are all the same in the DNA profile." The wine community, however, widely considers them separate varieties that produce wines of distinct flavor profiles. Tempranillo Blanco, a newly registered variety for Rioja, is also technically the same variety as Tempranillo. A major genetic mutation that produced this clone took out all the anthocyanin genes, and although not strictly a new variety, it is for all intents and purposes being treated as one.

Vine propagation occurs vegetatively, meaning that cuttings taken from vines are used to produce new plants, which are genetic clones of the parent variety. Attempts to grow vines from seeds are almost certainly doomed to failure because the genetic reassortment that takes place usually means the loss of positive features of the variety. Generally, growers are happy with the varieties they have and just want to improve them in ways that don't affect the expression of varietal character.

Within each of the different varieties a range of clones exist. Some of these differ through spontaneous bud mutations, which then result in genetically altered shoots. Almost always, such mutations are deleterious. Sometimes, however, they are positive, and can be propagated by cuttings taken from such an affected shoot, resulting in a new clone of the variety. Over enough time, a range of such clones might be developed. Other times, the clonal differences reflect nothing more than differing levels of virus infection, or perhaps epigenetic differences (heritable changes that aren't based on DNA sequence changes).

Vouillamoz adds, "The definition of clone is not really clear, either. For me, a clone is one or several mutations that are spectacular enough to the human eye for them to be selected. It is very subjective. For the producer, the color of the grape is important, or the size of the bunch is important. But maybe for a biochemist, a mutation in a biochemical synthetic pathway could be more important, but it is hidden."

Another mechanism underlying differences between varieties and clones is a process called chimerism. This is where separate portions of the plant have different genotypes. Mark Thomas and colleagues have demonstrated that the phenotype of the variety Pinot Meunier results from the interaction of two genetically distinct cell layers. They separated the two by tissue culture and showed that one layer is the same as Pinot Noir, while the other is a mutant that is insensitive to the plant hormone gibberellin, and produces a short stubby vine with a fruit cluster at every node, instead of the more usual mix of fruit clusters and tendrils. I asked Thomas how common this sort of grapevine chimerism is. "Since our work, other research groups from various countries have looked at other varieties and found similar results. So I would guess that most, if not all, old varieties would have accumulated somatic mutations [one occurring in any cell in the plant except the germ cells, the pollen or ovule] and a chimeric situation would be very common due to somatic mutations arising from one cell in a specific cell layer, and that mutated cell eventually taking over the whole layer."

Modern molecular biological techniques have provided some important new insights into the relationships among different grape varieties. Dr. Carole Meredith and her colleagues were the first to use microsatellite markers (also known as simple sequence repeats; a feature of complex genomes such as that of the grape that can be used as a molecular fingerprinting device) to sort out relationships between grape varieties that weren't apparent from the traditional chromosome studies, finding in 1997 that the parents of Cabernet Sauvignon were Cabernet Franc and Sauvignon Blanc—the first successful vine paternity test. Among other discoveries they have shown that Zinfandel is the same grape variety as Primitivo (grown in Italy) and Crljenak (from Croatia, also called Tribidrag there), and Chardonnay is the result of what was likely an accidental cross between Pinot Noir and an undistinguished white variety called Gouais Blanc.

"Microsatellites are pieces of DNA where the bases of the DNA repeat themselves. When you have coding DNA, the bases A, T, G, C are in a certain order that would translate into a protein," explains Vouillamoz, who has collaborated with Meredith in these studies. "At one point the DNA starts to stutter and to repeat, like GAGAGA one hundred times, for example. This part of the sequence does not contain any genetic message. We don't know what the use of these microsatellites is, but they exist in all organisms."

Vouillamoz continues, "These pieces of DNA are repetitions of bases that exist in every organism. In one single individual—either a grape variety or a human being—they are stable, and they differ between two individuals. If you look at, say ten different microsatellites, you get a DNA profile of your individual or grape variety. It remains the same within the grape variety, and it differs from one to the other. The first use of microsatellites [in grapes] was in 1993. The idea was to identify the grape varieties in the collections. Australians did it, and a few years later John Bowers at UC Davis (California) started to analyze the collection there. The idea was to ascertain which were duplicates: if two varieties are the same you can eliminate one from the collection and save money. Comparing the DNA profiles, he noticed that Cabernet Sauvignon had a profile very similar to Cabernet Franc and Sauvignon Blanc. So, he looked into the human genetics, to see how parentage testing was done. When he did the same with grape varieties, he realized that these markers were consistent with Cabernet Sauvignon being the progeny of Cabernet Franc and Sauvignon Blanc. He went on with the analysis and statistics, and 1997 was the first publication of grape parentage for Cabernet Sauvignon. This was groundbreaking news in the wine world as everyone thought all these varieties, especially in Bordeaux, had always existed. No: it was born some time, somewhere, and we know the parents. Most of the time they have disappeared or we know only one. Here we know both of them. No one would have thought that a white grape variety could be a parent of Cabernet Sauvignon, which is black berried. This was a breakthrough, and this is the technique I have used since then."

Vouillamoz says that this technique is quick and inexpensive, and he charges about $130 per sample. "When I was at UC Davis, one producer

Ancient wine: the new field of molecular archeology

The science of archeology must be a frustrating one at times. Detective work of the highest order is necessary to meld together the few remaining pieces of surviving evidence of a bygone era into some sort of coherent story. Patrick McGovern, an archeologist at the University of Pennsylvania, is probably the leading expert on the ancient origins of wine. But rather than just rely on old fragments of pottery and a few vine seeds, McGovern has turned to advanced molecular biological techniques to provide new evidence to shed light on the origins of wine in ancient civilizations. This new avenue of research has been dubbed "molecular archeology".

McGovern has collaborated with grapevine molecular biologists to study the DNA of ancient grape relics, such as seeds. Using similar techniques as those employed by Carole Meredith and her colleagues to assess relationships among modern-day varieties, McGovern is using microsatellite repeats in the ancient DNA to identify the grape variety and its relationship to modern vines. It is a work in progress as he and his collaborators continue to fine-tune the complex process of extracting useful DNA from ancient plant tissue and then making sense of the results. As yet, more concrete results have come from the array of chemical techniques that have been used to study residues present on archeological samples, including infrared and UV spectrometry, gas chromatography/mass spectrometry, and liquid chromatography/mass spectrometry. Together, these are powerful tools for providing scientifically reliable answers to questions that were previously just a matter of conjecture.

McGovern's DNA search for the site of domestication of the Eurasian grapevine, dubbed the Noah Hypothesis, has supported the thesis that it first happened in the Near East. As well as genetic evidence, linguistics and archeology point toward an area in what is known as the Fertile Crescent, in eastern Anatolia, as being the site of the first domestication of *Vitis vinifera*. However, Transcaucasia also remains a possibility.

I asked one of his collaborators, José Vouillamoz, how many domestication events there were. "We don't know how many. But our belief is that there were a small number of primary domestications somewhere in the Near East, most likely in southeastern Anatolia. It wasn't a single plant: we don't believe in an 'Eve' hypothesis, like we have for humans. I think we had a large population of wild grapevines, and from this large population a few hermaphroditic individuals gave birth to all the varieties that we have today. In a natural wild grapevine population, two to three percent of the plants are hermaphroditic. It is a small number. It has created a kind of bottleneck, genetically speaking, because they selected only the hermaphroditic plants, which was the starting point of grape domestication. I believe there were several primary domestications. I am still not convinced about secondary domestication centers in Western Europe, perhaps in Greece or Italy. I am open-minded about this, though. I would be more in favor of a secondary domestication center in the Iberian Peninsula, because they have a lot of wild grapevines."

wanted to plant one hundred hectares of Syrah, and he wasn't sure of the nursery. He sent us a sample and we analyzed it, and found out it wasn't Syrah, but Peloursin. It is in the same family as Syrah, but a very rare variety. It is a parent of Durif, which is a cross between Peloursin and Syrah, and is often called Petite Syrah. Carole Meredith analyzed the vineyards in California and nine out of ten times what they call Petite Syrah is Durif, and one in ten times it is Peloursin."

A famous example of mistaken identity was with the Australian Albariño, which was recently found to be Savagnin (shown by DNA analysis to be the same as Traminer, the nonaromatic version of Gewürztraminer). This mistake was revealed when France's top ampelographer Jean-Michel Boursiquot toured Australian vineyards and suggested that the vine they were calling Albariño was in fact Savagnin. Australian authorities compared samples of Spanish Albariño and French Savagnin with what they were selling as Albariño, and realized their mistake. From the 2009 vintage, Australia's "Albariño" producers have renamed their wines Savagnin Blanc or Traminer.

FOUNDER VARIETIES

Savagnin turns out to be very important, because it is what is known as a Founder grape variety. "We realized that a small number of grape varieties have given birth to all the diversity that we observe today," says Vouillamoz. "This is a new concept: we are the first ones to propose this hypothesis, because when you read other books or even scientific papers, they give the impression that all the important grape varieties that we have today, have been introduced a long time ago from different places—from Egypt, the Near East, the Middle East, and so on. I do think that we had a limited number of ancient introductions of the ancestors of what we have called the founder grape varieties, among which we find Savagnin Blanc, Pinot, and Gouais Blanc [the parent of no less than eighty varieties that are cultivated today in Western Europe]. We have established so far a total of thirteen founder varieties that gave birth to most of the diversity that we observe today in western Europe, but many more remain to be discovered in other regions were grape family trees have not been studied yet."

2 Terroir: how do soils and climate shape wines?

"Terroir" is a concept that is rapidly emerging as the unifying theory of fine wine. Once almost exclusively the preserve of the Old World, it's now a talking point in the New World, too. The traditional, Old World definition of terroir is quite a tricky one to tie down, but it can probably best be summed up as the way that the environment of the vineyard shapes the quality of the wine. It's a local flavor, the possession by a wine of a sense of place or "somewhereness." That is, a wine from a particular patch of ground expresses characteristics related to the physical environment in which the grapes are grown.

The goal of this chapter will be to give a broad introduction to this topic, examining why it is still a controversial issue. Then I'll focus on the scientific underpinnings of this concept, concentrating on teasing out the relationship between vineyard characteristics and wine flavor.

DEFINITIONS

One of the problems with many discussions of terroir is that this word means different things to different people. Indeed, defining terroir in precise terms is quite difficult, partly because it is a word used in three rather different ways.

A SENSE OF PLACE

The primary definition is that terroir is the possession by a wine of a sense of place. That is, the wine expresses flavor characteristics influenced by the properties of the vineyard or region it hails from. Immediately we see that scale is an issue here: environmental variation affects wine flavor, but this variation operates on a number of

scales. Wines made from grapes harvested from different parts of the same vineyard may well taste different. On the other hand, there might be characteristics held in common by wines made from larger geographic regions that are evident when these wines are compared with those of other regions, for example, Burgundian Pinot Noirs compared with Californian Pinot Noirs. And which factors should be included in the definition of terroir? It is clear that human intervention in terms of viticultural and winemaking practices may also confer a sense of place to a wine, but some wouldn't count the human element as part of terroir.

This raises the interesting question of the differences between terroir and typicity (the way a wine has a taste typical of the region or appellation). Could winemaking play a role in maintaining typicity? Certainly, in the classic Old World regions where terroir is so precisely delineated, the fact that winemakers commonly use similar techniques could help lend a distinctive regional style. Winemakers could also be adapting their techniques to best exhibit regional differences in their wines. This typicity, which owes more to human intervention than it does to classical definitions of terroir, is still of merit because it helps maintain a sort of stylistic regional identity. In general, though, terroir is easier to conceptualize if winemaking is excluded. "I don't see winemaking as part of terroir," says Jeffrey Grosset of Australia's Clare Valley, "but rather that poor winemaking can interfere with its expression and good winemaking can allow pure expression."

At Felton Road, the celebrated Central Otago (New Zealand) winery, terroir is seen as a partnership. Owner Nigel Greening explains that when it comes to blending the single-vineyard wines, Cornish Point and Calvert, only about 30–40 percent of the production will go to a single-vineyard bottling, with the balance going to the winery's Bannockburn label (the wines from other two single vineyards, blocks 3 and 5, almost all end up in the single-vineyard bottlings). "So, we have a dilemma," explains Greening: "we have to choose which 30–40 percent to use. There will typically be eight lots from each of the vineyards, and we will taste them about three times, blind. We score them not for their quality as a wine but for their expression of site. The Calvertiest ones become Calvert, and those that show the least Calvertness go into Bannockburn. The same applies to Cornish Point."

THE VINEYARD SITE

The next use of terroir is in describing the vineyard site itself: the combination of soils, subsoils, and climatic factors that affect the way that grapes grow on the vine and thus influence the taste of the wine made from them. This is probably the least controversial use of the term, because it is purely descriptive.

GOÛT DE TERROIR

Finally, there is a third use of this word. The term *goût de terroir* is sometimes used to describe flavors that are presumed to be imparted by the vineyard site itself. Thus someone might say that they taste notes of "terroir" in the wine. This is the most confusing use of this word, and in scientific terms it is hard to defend because it makes assumptions about mechanisms that can't be demonstrated.

TERROIR IN PRACTICE

Let's try to explain terroir in practical terms. Take a property with three different vineyard sites, one flat, one on a south-facing hillside, and one on a north-facing slope. We're assuming for the sake of simplicity that the same grape varieties and clones are used in all three, that the vineyards share the same geology, and that they are farmed the same way. Three wines are then made, one from each vineyard, in identical fashion. They will likely all taste different. That is terroir in action. Typically, vineyard sites will differ not only in one variable, as in this example, but in several. And such factors as differences in slope, orientation, or soil type will also influence decisions about which varieties are planted where, bringing further variables into play.

TERROIR IN THE NEW WORLD

One interesting question surrounds why there has recently been a transformation in attitude among New World winemakers, where for so long the job of grape growing was seen merely as a mundane prelude to the work of the all-powerful winemaker. Even fairly recently, the New World response to "terroir" was typically that it was a last-ditch marketing ploy by European winegrowers who were panicking about their increasing loss of market share. This turnaround has occurred for two reasons. First, New World wine growers realized that one of the keys to wine quality starts with grapes that show homogeneous (even) levels of ripeness, and also recognized the role that natural variation within and between their vineyards plays. With the increasing adoption of a technique known as precision viticulture, vineyards are commonly broken up into subplots sharing similar characteristics (known as natural or basic terroir units) so that vineyard interventions can be precisely targeted to where they are needed. The second reason is that New World winemakers have realized that regionality is the way forward with fine wine.

But you don't have to look too far below the surface to see that there are subtle but important differences between Old World and New World notions of terroir. Speaking generally, in the Old World terroirists aim to make wines that express the typicity of the specific vineyard site, whereas in the more pragmatic New World, understanding terroir is seen as a route to improved quality. Of course, there are exceptions to this generalization.

WINEMAKERS' VIEWS ON TERROIR

Although the idea that soils and climate influence wine flavor seems to be rather obvious, terroir is in fact a highly controversial concept. It doesn't help that it's long been regarded as the exclusive preserve of the French. "The French feel they have ownership of 'terroir'," says Barossa winegrower Charlie Melton, "but in good winemaking the idea is universal." Melton prefers not to use the "T" word itself, but instead talks of characterization of distinct vineyard blocks. "In the Barossa the subregions have their own character," he says. "Wines from the southern end have finer, slightly sweeter aromatics, and those from the northern end have a *garrigue*-like earthiness rather than sweetness." He points out that the humidity is on average three percent higher in the southern end, and that the soils in the Barossa vary quite widely, with adjacent blocks sometimes making quite different wines. "Call it terroir if you like," he adds.

OBJECTIONS TO TERROIR

But one vigneron of the New World who objects to the notion of terroir is Sean Thackrey, a Californian winemaker famous for his single-vineyard Orion wine. "My objection is simply that it's so ruthlessly misused, and with such horrifying hypocrisy," says Thackrey. "It's very true that fruit grown in different places tastes different. In fact, it's a banality, so why exactly all this excess insistence?" Thackrey himself allows terroir to influence his work, claiming, "I don't know how it will be possible to observe the delicacies of change in a particular vineyard more attentively than I do in making the Orion." But he still feels that the French overemphasize terroir for largely economic motives. He describes it as "an intensely desirable and bankable proposition because their property

can then be sold, transferred, and inherited with the full value of the wine produced from its grapes attributable to the property itself." Thus the work of the winemaker and viticulturalist is played down and the role of the vineyard site talked up.

A KEY TO SUCCESSFUL BLENDING

Portugal's Dirk Niepoort, well known for his trailblazing Douro table wines, adds a slightly different complexion to the debate: "As a rule, I believe blended wines to be better than single vineyard wines." This doesn't mean he's a nonbeliever, though. "I believe that terroir is essential," he says. "I think that a good blender has to be someone who understands, knows, and respects their different terroirs." Niepoort maintains that the search for great terroirs is very important to him. "But it is not only finding the terroirs that is important: it is also important to understand them and then adapt your winemaking to them."

CLIMATE VERSUS SOILS

Climate is clearly vital to the concept of terroir. The average climate of a vineyard site determines which grapes it can effectively grow. In large vine repositories, such as the ampelographic collection at the École Nationale Supérieur d'Agronomie de Montpellier (National Superior School of Agronomy of Montpellier), in France, the earliest and latest varieties ripen some two months apart. Remember: these are all varieties of the same species. Vines are indeed incredibly fussy about climate, and only perform well if fairly narrow climate parameters are met.

But while climate determines which varieties can be grown successfully at a particular location, it is the soils that are of real interest to wine science types. If the climate is just right for your variety, this is but part of the story, as one of France's most famous wine region, Burgundy, illustrates perfectly. There aren't that many places where Pinot Noir can be grown successfully, because among all grape varieties, it is the fussiest of the fussy. It has taken wine producers in the New World decades to find sites where they can get it to perform properly. Even within Burgundy, though, some vineyards do brilliantly with it while others, just a few yards away, make mediocre wine. The climate isn't changing much as we progress from regional to village to Premier Cru to Grand Cru vineyard; it is the soils and geology that are making the difference.

Burgundy is therefore seen as the test case of terroir. It is a region divided into a patchwork of vineyards based on long experience. Over hundreds of years, people observed that certain vineyards did consistently better than others year after year. This resulted in a hierarchical classification and structuring of vineyards based on differences in wine characteristics, and these characteristics have since been found to have their origin in the vineyard's physical properties. When the boundaries for the Burgundy vineyards were put in place, no one knew much about geology and soil maps: they just recognized that some vineyards, rather inexplicably, were privileged, and produced better wines. But now geologists can show that the hierarchy of Burgundy vineyards reflects changes in the subsoil properties that influence grape and thus wine quality.

MECHANISMS OF TERROIR: A TASTE OF THE SOIL?

The notion of terroir is fundamental to the wine industries of such Old World countries as France, Italy, and Germany. It's a philosophical framework within which winegrowers work. Local wine laws are built around the concepts of appellations, which lend official sanction to the idea that a combination of certain vineyard sites and grape varieties creates unique wines that faithfully express their geographical origins. Correspondingly, many Old World growers feel they have a duty to make wines faithful to the vineyard sites they are working with. These

growers will commonly make associations between properties of the wines and the soil types the grapes are grown on. In some cases these putative associations are quite specific. People will talk about mineral characters in wines and associate them with the minerals in the vineyard, taken up by the roots of the vines. Do chalk, flint, or slate soils impart chalky, flinty, or slatelike characters to wine? As a scientist who has a working knowledge of plant physiology, I find this notion, which I call the "literalist" theory of terroir, implausible. Yet I can't get away from the fact that an overwhelming majority of the world's most compelling and complex wines are made by people who hold the notion of terroir as being critical to wine quality.

Thus one of the goals of this chapter is to explore the mechanisms of terroir, focusing specifically on soils. Just how do soils affect wine quality? Is it a direct or indirect relationship? What are the scientific explanations for terroir effects?

If we are going to frame terroir in scientific language, then we'll need to start with some plant physiology. The miracle of the plant kingdom is that these complex organisms build themselves from virtually nothing: all that a plant needs to grow is some water, sunlight, air, and a mixture of trace elements and nutrients. All the complex structure and chemistry of an oak tree, a daffodil, or a grapevine is fashioned from these very basic starting ingredients. What do vine roots take up from the soil? Primarily water, along with dissolved mineral ions. It seems implausible that such a complex structure as a vine is created from virtually nothing by photosynthesis—the capture of light by specialized organelles called chloroplasts, which turn light energy into chemical energy that the plant can use—but that's the way it is. As Richard Smart emphasizes, "All flavor compounds are synthesized in the vine, made from organic molecules derived from photosynthesis ultimately, and inorganic ions taken up from the soil." Professor Jean-Claude Davidian of the École Nationale Supérieure d'Agronomique in

Above Alluvial soils in Margaux, in the southern Médoc in Bordeaux. These are stony and free draining.

Above Basaltic soils in the Walla Walla Valley that straddle Washington State and the border of Oregon.

Above Classic clay and limestone soils (*argile calcaire*), Bourgogne.

Above Albariza soils from Jerez, Spain. These prized soils are found throughout the Sherry triangle. There are actually six different types, all based around limestone and clay.

Above Chablis is famous for its limestone-based soils, though there's a little variation around the region. Here's a soil with small limestone pebbles in clay-rich loam.

Above Granitic soils in Fleurie, Beaujolais.

Above Alluvial pebble soils in Toro, Spain.

Above Granite soils, Grand Cru Brand, in Alsace.

Montpellier echoes these sentiments. "Nobody has been objectively able to show any links between the soil mineral composition and the flavor or fragrance of the wines," he says. Davidian adds, "Those who claim to have shown these links are not scientifically reliable."

TREATING VINES MEAN TO KEEP THEM KEEN

It is helpful to think about plants as sophisticated environmental computers. In the same way that we sense the world around us and then use this information to guide our actions, so do plants, but over a longer timescale. Literally rooted to the spot, they adapt their growth form to best suit the local conditions. This extends to their reproductive strategies. Generally (and simplistically) speaking, if conditions are good, then plants opt for vegetative growth; if they are bad, they choose to reproduce sexually, which means fruit production. So viticulturalists want to treat their vines mean enough that they focus on fruit production, while giving them just enough of what they need so that they don't suffer from water or mineral deficit, which would hamper their efforts at producing ripe fruit. Thus many viticultural interventions aim at encouraging the vine to partition nutrients to the grapes so that they ripen properly, rather than concentrating on growing more leaves and stems (vegetative growth or "vigor").

An example of this "environmental computing" is seen in the growth of plant roots. Root growth is determined by interplay between the developmental program of the plant and the distribution of mineral nutrients in the soil. The roots grow to seek out the water and nutrients in the soil. To do this, it appears they sense where the various nutrients are and then preferentially send out lateral shoots into these areas. Low levels of nutrients in the upper layers of the soil results in the roots growing down to a greater depth, which is likely to improve the regularity of water supply to the vine. "Vines have roots which can reach up to three meters [nine feet] in depth," reports Davidian. "These deep roots can actively take up water and minerals, even though most mineral ions are more abundant at the root surface."

HOW SOILS HAVE THEIR EFFECT

Soils differ in their chemical and physical properties. Many scientists consider that the latter are more important for terroir effects. This was demonstrated in the pioneering work of French scientist Gérard Seguin, who conducted a survey of the properties of the soils in the Bordeaux region in the mid-1980s. Seguin couldn't find any reliable link between the chemical composition of the soil and wine character or quality, and maintained that it was the drainage properties of the soil affecting the availability of water that mattered. He concluded that it is "impossible to establish any correlation between the quality of the wine and the soil content of any nutritive element, be it potassium, phosphorus, or any other oligo-element." The verdict was that it was the physical properties of the soils, regulating the water supply to the vine, that were all important in determining wine quality. The best terroirs were the ones where the soils are free draining, with the water tables high enough to ensure a regular supply of water to the vine roots, which then recedes on *veraison* (when the berries change color) so that vegetative growth stops and the vine concentrates its energies on fruit ripening.

SOIL CHEMISTRY EFFECTS

But before we give up on soil chemistry as an important factor in terroir, it's worth taking a look at the effects of mineral nutrition on plant physiology. I spoke to a number of researchers who are actively working on plant mineral nutrition, to see whether their work might shed some light on the mechanisms of terroir transduction. "I wouldn't be at all surprised if soil chemistry had an effect on the expression of genes that are involved in the production

Above Sandy soils in Franschhoek, South Africa.

Above Sandy granitic soils in Portugal's Dão wine region.

Above Gravelly, free-draining soils in the Gimblett Gravels district of Hawke's Bay, New Zealand.

Above Schist soils in the Douro Valley, Portugal.

Above Schist soils in New Zealand's Central Otago wine region.

Above Schist soils in the Mosel, Germany.

Above Loess soils, Washington State.

Above Red clay soils, Rheinhessen, Germany.

of the compounds that determine flavor," says Professor Brian Forde of Lancaster University, in the UK. "There is certainly plenty of evidence that plants are tuned to detect and respond to soil nutrients," he adds. "The balance between the nutrients (nitrogen, phosphorus, potassium, sulfur, and calcium, and even the micronutrients) is likely to be important and the plant stress responses elicited by limiting amounts of one nutrient would probably be subtly different from the stress responses elicited if another nutrient is limiting." Forde referred me to some publications showing that the levels of various plant metabolites were significantly altered under different nutrient regimes. At a more detailed level, it is now clear that patterns of gene expression in plants are altered by the presence and absence of various nutrients.

And then there is the effect of nutrition on yeasts in fermentation. The composition of grapes determines the composition of the must, and as yeasts are producing many of the aromatic compounds in wine, and their metabolism is highly dependent on their nutritional state, this is another way that site can alter wine flavor, albeit through two biological intermediaries—the grape berry and the microbes doing fermentation.

I spoke to Professor Malcolm Bennett and Dr. Martin Broadley of Nottingham University, in the UK, who have studied the effects of phosphate deficiency on plant gene expression. Broadley feels that it won't be too long before we have a much better idea about the influence of soils on wine flavor. "There is a large amount of work underway to understand the molecular biology of grapes, and scientists are identifying genes that influence wine flavor," he explained. "As more grape molecular biology is known, the easier it will be to understand mechanisms of terroir on wine taste. When genes encoding for proteins that influence wine taste are identified, then the effects of different components of terroir (i.e. the availability of different minerals, soil

pH, soil water content) on specific biochemical pathways can be identified and tested. This research may allow current agronomical practices to be improved to enable better-tasting grapes to be produced, or it might even allow varieties of grapes to be selected or bred more effectively."

Even if science leaves us with what currently looks like rather an emasculated version of terroir, I don't think that this necessarily diminishes the importance of this cherished concept. Wine growers who use terroir as their guiding philosophical framework and focus on the importance of the soil are responsible for a disproportionately large share of the world's most interesting wines. Even though it seems that there is unlikely to be a direct link between soils and wine flavor, by framing their activities within the context of a soil-focused worldview and trying to get a touch of "somewhereness" and minerality into their wines, winegrowers might be vastly increasing their chances of making interesting wine. And that is something the world needs more of.

As with so many other areas of viticulture and winemaking, it seems that more research is needed. While the current consensus is that it's the physical rather than the chemical properties of the soil that are all important in separating good terroirs from bad ones, there's still the possibility that new data could emerge demonstrating that soil chemical properties can modulate grape characteristics significantly through altering gene expression.

What is the use of this research, besides satisfying our intellectual curiosity? In an ideal world we could all drink fine wines, such as first-growth Bordeaux and Grand Cru Burgundy, every day. We can't because there isn't much made, and they are expensive. It is all very well if you own some of Montrachet, or have vines on the hill at Hermitage, or preside over a Médoc First Growth. You are blessed. Winegrowers with less auspicious terroirs, however, would understandably like to make better wine, and if scientists uncover precisely what it is that makes a great vineyard as opposed to an ordinary one, then they might be able to implement better management strategies that would improve their wines, or engineer vines that perform superbly on indifferent terroirs. After all, there's nothing magical about great terroirs: they are just patches of ground that naturally possess the conditions that encourage grapevines to produce grapes that, when handled correctly in the winery, can make outstanding wines. Wouldn't it be great if top-class wines were affordable for all? What a great scientific objective!

In the next chapter, I will be taking a closer look at the topic of soils, seeing how they might be important in shaping wine quality. I will also be addressing the issue of minerality, which is currently a hot topic in the world of wine.

Linking terroir and wine aroma: the chemical basis for the concept of cru

Maurizio Ugliano is a well-traveled wine scientist who works at the Università di Verona, in Italy. He has been involved in an interesting research project with the Tedeschi winery. "We started with this idea of linking wine geographical origin with aroma composition," says Ugliano. "We were thinking around the concept of *cru* vineyards: vineyards that because of anecdotal observation seem to have some elements that come back year after year." With his Ph.D student, he wanted to see whether specific vineyards had aroma fingerprints that were independent of vintage. "We know that if you change the location of the vineyard the wine will be different, but this was more about if we go back to samples from different years from that vineyard, do we see some recurring patterns that we could claim that are part of this notion of *cru*?"

They looked at five vineyards owned by Tedeschi, two historical vineyards that are in the Valpolicella Classico area, and three in the Valpolicella Orientali, which was added to the appellation more recently. They worked in real-life conditions, harvesting the grapes when the winery harvests them, for example. "We did not force harvest-date decisions, insisting that they all need to be at the same Brix," he says. "This is a strong point, because it means the observations we made are more likely to reflect real-life situations."

The three vineyards in the Valpolicella Orientali are within the same estate (Anfiteatro, Bàrila, and Impervio vineyards of Tenuta di Maternigo), so they are very close. "One part of the work focused on where vineyards are different at a microscale, looking at vineyards one next to the other," says Ugliano. "For three consecutive vintages we went back to these vineyards to rows that we had marked. We got the grapes and made the wines. We made two versions: one was a conventional Valpolicella Classico with no withering, and the other was in the style of Amarone [the grapes were dried for some time before vinifying]. We did this all in an experimental setting with replicates."

"In the first vintage we saw that the wines were quite different. There's nothing new in that. We observed a couple of interesting things in this first vintage. First, we realized that these grapes, which are red grapes, are quite rich in terpenes. We'd seen this in previous years before we started this experiment. But this is not only the linalool floral-like terpenes, which we have quite a bit of in the young wines but with a few years of aging go away, but also other terpenes with menthol/eucalypt aromas. In the tasting notes we find that often tasters report these aromas in these wines."

"Because this diversity was already quite clear in the first year, we thought it would be mostly associated with grape-derived compounds like terpenes and norisoprenoids and sesquiterpenes. But we also saw the fermentation-derived compounds changing quite a bit depending on where the grapes were coming from. We went back to look at this from a yeast perspective. Part of this diversity that is associated with the vineyard is modulated by what the yeast is doing." Some of this diversity is also associated with compounds that the yeast is making *de novo*: things that are not in the grape, which the yeast makes. "These are esters and higher alcohols that are associated with the site," he says. "In the grape composition, there are some elements that are triggering the same yeasts to do different things."

There are two levels to this *cru* aroma signature.

"One is the stuff that you have in the aroma compounds and precursors in the grapes. These are influenced by the way the grapes are ripening, and so on. The other is the environment that the grapes create around the yeast that triggers the yeast to do different things."

They repeated these experiments for three vintages. "We were able to demonstrate that there are certain vineyards—three out of the five we studied—where every vintage is being characterized by a certain type of aroma profile," says Ugliano. "For example, there is one vineyard that systematically gives more terpenes in the wine than all the others. This is quite tricky to see because the levels of terpenes change with the vintage: there are some vintages that are rich in these compounds, and others less rich, across the board. Then if you look at this from a relative point of view within the same vintage, you find these aroma signatures of the vineyard."

"Vintage is certainly a major element in determining whether a wine will express certain characters, most likely because of the climatic characters of the vintage. But then within that trend, the aroma signature associated with the vintage was very clear. We could only see these after three years of study, but once we got all the data it was clear that this concept really exists."

To those who taste widely, it is not big news that different vineyards have characteristic aroma signatures. But this is one of the few successful attempts to nail down the chemical signature. "From your perspective where you are able to attend many tastings and compare wines, you probably stumble across characteristic aroma signatures frequently,"

says Ugliano to me. "From a purely chemical and analytical point of view, when no information about the winery/vineyard is provided it allows us to capture the chemical bases of these differences."

"In addition to the more usual grape aroma compounds that we would expect to mark these aroma signatures, why were the fermentation derived aromas also associated with the vineyard?" he asks. "We found that the nitrogen in the grapes that acts as a yeast nutrient plays a big role in this. This brings us back to the fertility of the soil. Because we were measuring the amount of nitrogen in the grapes every year we were able to correlate the amount of esters, for example, in the wines with the amount of nitrogen in the soil. This is interesting because when we think about soil we think about the minerals or the microelements, such as iron. They certainly have a role to play, but here we found that it was the nitrogen in the grapes that was important, and this is an important marker of the plant/soil interactions."

"Wineries tend to add nitrogen during fermentation in order to assist the yeast. One implication of what we observed was that if you do that, you will be less likely to express the link between the wine aroma and the vineyard. You are masking one of the elements that drives the expression of this link."

In a final experiment, they took the grapes from two of the five vineyards—the ones that were more extreme in terms of diversity. "Each of the batches of grapes was inoculated with four commercial yeasts plus one was a spontaneous fermentation. Then we did sensory and chemical analysis on the wines. Our key question in terms of interpreting data was whether the greatest diversity was associated with the origin of the two batches of grapes or with the type of yeast used for fermentation. At every level—in terms of both sensory and chemistry—by far the biggest impact was due to the grape origin. The yeast was quite minor in terms of impact, including in the spontaneous fermentation. It was the grape origin that was driving the diversity that we could see."

"For these wines we can now say there is a chemical basis for the *cru* concept," says Ugliano.

Cavit's PICA: a powerful vineyard model used to guide decision-making in Trentino's vineyards

No vineyard is completely consistent in terms of its physical properties. However, the technology exists to enable researchers to map vineyard variation, for example, by surveying the soil and subsoil, and then to respond to this. You can start by making a map using electroconductivity, and then decide, on the basis of these results, where to take core samples and observe what the actual soil profile looks like. Then on the basis of this information, you can dig soil pits at various points. This will give you information about the physical characteristics of your vineyard soils and subsoils.

Next you might want to put data loggers in various parts of the vineyard, to see how the microclimate varies. Some spots might be warmer, others cooler. The climate in the vineyard might differ from the local weather station that gives long-term data for the region, and this will be useful to know.

And you can track how these physical characteristics affect vine growth. Remote sensing, made much cheaper by the advent of drones, can give you NVDI (normalized difference vegetation index) readings that show where the more or less vigorous spots are. You can then relate this to the physical characteristics. Assuming the vineyard Is planted with the same variety and rootstock, and was planted at the same time, then trunk diameter readings will also reveal information about vigor. Or you could drive through and record canopy density and then use GPS to provide a vineyard map. Or if you machine-pick, then yield can be mapped with GPS too.

The challenge is to then combine all this information in a single model. If data over several seasons is compared, and then combined with weather data and outcomes such as disease incidence, yield, and grape-quality parameters, suddenly you have a very powerful viticultural model. This can be used predictively to assist decision-making. The model data can be combined with information about disease or pest biology, fed with weather forecast information as well as actual weather readings, and the readout can tell you whether you need to act or not, and how.

Feed in information on grape phenology, grape-variety-specific data, and maturity analysis on the site, and suddenly you have a great tool for tracking maturity that can be ground-truthed and refined at certain points in the growing season. It can tell you when to sample, and reduce the need for continuous monitoring, which can be time-consuming if you are running several vineyards at various locations.

With these thoughts in mind, I was very interested when I got the chance to visit Cavit, a large cooperative winery in Trentino, Italy. They have ten wineries spread around the Trento area, and in addition have a network of 4,500 growers, who together farm 5,400 hectares through the region (altogether Cavit source from some 7,000 hectares). Keeping track of this many vineyards is an enormous task, but they have founded something groundbreaking: a data-driven solution called PICA (Piattaforma Integrata Cartografica Agri-vitivinicola).

Cavit have developed a sophisticated model mapping data from all the vineyards they source from, and integrating this with real-time weather data. Their goal was to develop a data-based tool that allowed them to improve quality and preserve the environment, in order to be sustainable in the long term.

Above A screenshot of Cavit's PICA in action.

PICA is the result of a five-year research project by Cavit in collaboration with MPA Solutions (an IT company which also worked on the development of the traffic-light system in Los Angeles) and two leading research centers in Trentino: the Edmund Mach Foundation (FEM) and the Bruno Kessler Foundation (FBK).

The software is the property of Cavit, and it maps 60 percent of the Trentino territory. It includes all 5,400 growers. The benefits include soil classification (they know what kinds of soils they have in each vineyard), which will also help guide future planting, in conjunction with information about the climate and exposure.

PICA has now been running for nine years and it is delivering results. It helps plan irrigation, incorporates geographic, weather and climate studies, and allows defense-monitoring of pathogens. The data also takes into account the training system and chooses the best one for the variety and the kind of wine they would like to have. Not all the growers are technology friendly, but there is some training available to them. The result is a continuous flow of information between the Cavit team and these growers.

Additionally, PICA can control and monitor the right time for harvest, forecasting when the grapes will be at optimum ripeness. This particular element is called "Harvassist".

"Our DNA is delivering quality at different price levels, and PICA helps a lot," says Andrea Faustini, who is one of the oenologists and agronomists in the Cavit team. "When you produce sixty million bottles, we are proud to say that we deliver quality at different price levels, and this is a fantastic tool to support us so we can do this."

The day before my visit, there was some rain. "When it rains and there is the right condition for disease, we send the growers a message," says Faustini. "For example, we might say if the forecast is good for the next day, don't do a treatment, wait and see." It is a tailor-made approach, specifically targeted. "We send SMS messages—quick updates—a couple of times a week," he says. "We also send a bulletin out and the growers can read the message." If they advise treatment, it will give different strategies, depending on whether the grower is organic, sustainable, or conventional, outlining the options available.

3 Soils and vines

To the ancients, the idea that plants are formed from the soil would seem self-evident. The communion between the roots and the earth suggests that the composition of plants and, by extension, the fruit they produce, is determined largely by the composition of the soil. Modern science, however, paints a rather different picture. The fact that you can produce perfectly delicious-tasting fruit using hydroponics, where the plant is supplied with just water, light, and a solution of 16 trace elements, demonstrates that the intuitive notion that the soil makes the plant is quite mistaken.

Plants use light, water, air, and trace elements to synthesize everything they need. Think of them as remarkable chemical factories, taking very basic raw ingredients, and synthesizing complexity from them. Moving to viticulture, specifically, grapes—the starting place for wine production—are made entirely through the process of photosynthesis and the subsequent biochemistry that builds and modifies the building blocks of sugars into complex biology. The soil? It is merely supplying water and dissolved mineral ions. These nutrient minerals are derived from the vineyard geology, but they are only needed in only tiny quantities by the vine and have little if any aroma or taste. But let us leave scientific fundamentalism aside for a second, and consider the experience of winegrowers worldwide, over many centuries. This experience testifies that soils are actually vital for wine. You can take a trip to a vineyard region, such as Burgundy, and discover that when vineyard boundaries are coupled with changes in underlying soil structure, two neighboring vineyards can differ significantly in the quality of wine produced.

That underlying geology impacts wine so strongly is undoubted. The cost differential between a *grand cru* Burgundy and a lowly generic Bourgogne, or even a respectable village-level wine, is such that there is a significant financial incentive for a winegrower to do all they can to improve the quality of their wine. But even where great care is taken in the vineyard, yields are dropped, and the highest level of winemaking is practiced, there seems to be a quality ceiling that is imposed by the vineyard. So, we have a dilemma to solve: how is it that soils seem to be so important for wine quality, when science indicates that they are only playing a limited role in influencing the flavor of grapes? This is one of the questions that intrigue me most in the world of wine, and I am going to attempt to answer it.

ROCKS, STONES, AND SOILS

Those of us in the wine trade do tend to talk quite a lot about the geology of vineyards, so it is helpful to explore some of the science behind what we're talking about. So how does a rock become soil? Very slowly is the simple answer. But in geology, time is not of the essence. Slow is fast enough. The two processes that need to occur are weathering and the incorporation of organic material.

Consider a rock exposed to the atmosphere. It is going to get rained on, and it will experience temperature changes. Rain is very slightly acidic, because atmospheric carbon dioxide is dissolved in it. And, if any water should get into cracks or microfissures in the rock, and then freeze, because water expands by nine percent on freezing, it will exert pressure on the rock. These forces may be small, but over time, they will have an effect.

Organisms also play a role. Primary colonizers, such as bacteria, algae, and lichens dramatically accelerate the process of weathering. Lichens are amazing. They are a fusion of two quite different organisms, a fungus and an alga. The fungus is the host: it breaks down the surface of the rock by acidic secretions, releasing minerals that can then be utilized by the alga for photosynthesis, which produces carbohydrates that are then shared by the fungus. Lichens are thus able to grow where very little else can, such as on the surface of a rock. Lichens are resistant to drying out, and actually only grow when they are in the process of being wetted or are drying. Mosses are also effective colonizers, able to survive repeated cycles of wetting and drying, and needing very little mineral input to grow.

As the first colonizers of rock become established, the first elements of soil begin to be developed, albeit on a basic scale. Organic material from the decomposition of organisms begins to be incorporated with small particles of rock produced by mechanical and chemical weathering, and a succession of plant species then begins.

So, this is the simplest sort of soil: parental bedrock undergoes weathering, and organic material is combined with the weathered particles. But soils can also move around a little, especially when water is involved. There's the action of glaciers, carving through the landscape and scattering material around. Rivers also move material, depositing it. For example, the course of rivers might change quite a lot, creating alluvial fans where the deposits show an irregular distribution on a valley floor. Wind can also shift soils around: "loess" is the term used for wind-blown soil.

Of course, there are also the more ancient geological movements that need to be considered, such as continental drift. Mountain ranges getting pushed up, seabeds rising up (or sea levels falling), crumpling and folding of the Earth's surface, volcanic activity—all these will alter the way soils develop in complex ways. Generally speaking, soils themselves consist of particles of different sizes. There are rocks and stones, but the interesting parts are the three different particle sizes known as sand, silt, and clay. The size limits for these differ with classification type but, generally speaking, coarse sand is 2–0.2mm, fine sand is 0.2–0.02mm, silt is 20–2μm (micrometer), and clay is less than 2μm.

Add humus, an organic material to the mixture, and you have your soil. Humus is mostly plant residues that have been mineralized and recycled by soil organisms. But, in addition to this, there is ongoing biological activity because soils are teeming with life. The organic material in soil is vital for its structure and texture, both of which are important soil properties.

The term "structure" describes the way that the different particles are bonded together into what are known as aggregates. This structure is a vital property, because if the aggregates are unstable, then the soil is easily compacted, and this can lead to poor water exchange with the roots and poor gas exchange (air is an important component of healthy soils).

The stability of soil aggregates can only be maintained if the organic material—humus—that helps hold them together, is replenished. Vine roots need at least ten-percent air-filled space in the soil, and preferably fifteen percent, otherwise waterlogging is a problem. Dormant vine roots can withstand several weeks of waterlogging, but during the growing season even five days of waterlogging can start to kill them.

Crusting can occur at the surface of the soil if the aggregates there are broken down. This leads to the release of small particles, which are washed into the pores at the surface of the soil, making water penetration of the soil difficult. Structure is very important to soil because this will govern how

Above Vine roots have become exposed in a pit dug in a vineyard in Gualtallary, in Mendoza, Argentina. These stony soils are quite arid, and a deposit of calcium carbonate, known as pedogenic lime, often forms on the stones, or even in the soils themselves.

much space there is in the soil for air, as well as its ability to drain and retain water. And air, of course, is important in soil for encouraging microbial life.

Without structure, soils easily become compacted, reducing the flow of air. They are also more likely to become waterlogged. Soil microbes have been shown to be highly important in helping to form soil structure, in particular because of their role in stability and aggregate formation.

The role of microorganisms in the stability of soils has been studied in depth by Claire Chenu—one of the leading researchers on soil microbes and their role in soil quality—and her colleagues. A paper they published in 2011 reports on seven long-term experiments on different soil plots. They found a rapid increase in soil aggregate stability when practices that increase soil carbon (organic content) were implemented, such as no-till agriculture, the permanent cover of the soil by plants, and repeated compost additions. Follow-up laboratory experiments showed this increase in stability was an indirect effect, occurring through the stimulation of organisms that decompose organic material, which helped aggregate the soil.

Another vital aspect of soils is known as "colloids," represented by clays and humus,

which are able to act as chemical stores, soaking up and releasing mineral ions. Clays have an enormous surface area to volume ratio, which makes them very good at this. There are also different sorts of clays, and the type of clay can be significant for viticulture.

Clays are composed of layers of silica and aluminum, but the actual structure can vary. In simplified form, clays can be separated into two different groups, called kaolinite/illite and smectite/montmorillonite. Kaolinite/illite clays do not swell much when they are wetted, and have a low "cation exchange" capacity (meaning, they have limited ability to hold nutrients and exchange them with plant roots). Smectite/montmorillonite clays swell more when wetted and then crack when they dry out. They are stickier and more plastic and have a high cation-exchange capacity.

A "clay" soil isn't composed solely of clay, but consists of different proportions of other particles combined with clay. The proportion of clay is important: a high clay content will allow the soil to store more mineral ions, and then exchange them with the roots, especially if the clay is smectite rather than kaolinite/illite. We'll return to clays later, because they are very interesting from a viticultural point of view.

WHAT DO VINES NEED FROM THE SOIL?

So, what do plants—and more specifically, vines—require of soils to be able to function? Surprisingly little, it seems. The key factor, and this can be a deal-breaker, is, unsurprisingly, water. Water availability and grape quality are inextricably linked to the point that some have suggested that the main significance of vineyard soils is in the water supply to the vine. Aside from water, vines need mineral nutrients, which are divided into macro- and micronutrients, depending on how much is required by the vine. (Bear in mind that they are able to get carbon dioxide and oxygen from the air, and hydrogen from water: these are, of course, essential.)

Nutrients needed for plant growth

Macronutrients

Nitrogen

Phosphorus

Potassium

Calcium

Sulfur

Magnesium

Silicon

Micronutrients

Boron

Chlorine

Iron

Manganese

Zinc

Copper

Molybdenum

Nickel

Selenium

Sodium

Nitrogen is the key nutrient, needed in relatively large quantities. But too much nitrogen is actually a problem for grape quality: it promotes vigor (the excessive growth of the vine and canopy). While the vine likes nitrogen, if it gets too much it forgets to produce top-quality grapes. And as an aside, it is worth mentioning that plants generally have a reproductive strategy where they favor vegetative growth when things are going well (they are clearly in a good place), and sexual reproduction when things are tricky (time to get out of here).

The extra shading of the lush canopy prevents exposure of the base of the canes (the part of the woody stem that the leaves and grapes form along) to light, which compromises fruitfulness for the following season. This is where next year's buds are being formed, and the vine doesn't want to produce fruit inside a dense canopy (remember, the vine is designed as a woodland climber), and this is a way of making sure that the fruit is produced near the edge of the canopy, where birds are more likely to find it. Overly vigorous vines with dense canopies also have more problems with disease, and if the grapes are shaded too much they can struggle to ripen. It also increases the amount of vineyard labor trying to keep the canopies in check. For high-quality grape production, vines actually prefer less fertile soils, with lower nitrogen content, especially in cooler climates. In warmer climates, water stress can reduce vigor naturally, so in those cases more fertile soils might not be as bad. If soils are too infertile and contain very little nitrogen, then this can cause problems for fermentation, because nitrogen in a form that yeasts can use (known as YAN, for yeast-available nitrogen) is needed for fermentation to carry on successfully.

It is becoming apparent that white and red grape varieties might differ in their nitrogen requirements. In an interesting study for his Ph.D thesis, Jean-Sebastien Raynaud looked at the influence of different soil types on the Doral grape variety grown in 13 different vineyards in Vaud, Switzerland, where the soil was reduced to the only variable. He found that soil nitrogen was a critical variable. The 13 vineyards could be grouped into two types of soil, and one type was much lower in nitrogen than the other. Low vine nitrogen resulted in higher soluble solids content in the wine, low malic acid, higher pH, and smaller berries. The lower vine and juice nitrogen resulted in the lowest-quality wines, with reduced aromatic complexity. But Doral is a white grape variety, so it seems that many of the factors that are associated with low soil nitrogen might actually be correlated with higher red wine quality.

Potassium is another important nutrient. A deficiency can reduce yields and fruit quality, as well as increasing the susceptibility of the vine to disease. Too much potassium can be a problem, though, especially for red wines. This is because most of the potassium is in the skins, and during red-wine fermentation will find its way into the wine. This can cause loss of acidity, because the potassium ions bind with tartaric acid, the main grape acid, and the combination precipitates out as potassium bitartarate. Thus acidity is lost, and that is almost always a bad thing in winemaking.

Calcium and magnesium matter, too. Apparently, the ratio of calcium to magnesium determines the tightness of structure of a soil. More calcium makes the soil structure looser. This is a good thing, because the soil will have greater air penetration (encouraging the life of the soil) and will drain more easily.

VINE ROOTS IN ACTION

"The extent, health, and physical and chemical environment of the roots must be a major key to the best ripening and terroir expression." John Gladstones, *Wine, terroir, and climate change* (Wakefield Press, 2011)

Vine roots respond to the conditions of the soils they are growing in. First of all, a large permanent framework of roots is established, followed by a network of finer lateral roots, and finally even finer tertiary roots, which are vital for uptake of water and nutrients. Nutrient uptake by the roots can be both passive and active. As the vine takes up water, it will usually take up whatever is dissolved in that water. But if the vine lacks specific nutrients, it can take them up actively, if they are present in the soil. There are some situations where the vine is fooled, though, by mineral ions that look quite similar, such that a deficiency of one can occur when there's an abundance of another. And, for example, in soils with a lot of limestone, chlorosis can be a problem. This is because in limestone-rich soils, vines find it very difficult to take up iron, which is needed

for photosynthesis, as a vital component of the green-colored chlorophyll pigment. As a result, the leaves turn a yellow color and are diminished in their ability to carry out photosynthesis.

A special layer of material, called the Casparian strip, surrounds the root endodermis (the layer of cells that circle the vascular tissue). This strip contains suberin, a waxy, rubbery material that is impermeable to water. Thus water and solutes entering the roots have to pass through plasmodesmata (pores in the cell walls) and therefore through the cytoplasm of root cells, before they can be transmitted to the rest of the plant. This gives the vine a level of control over what is taken up. The plasmodesmata are significant because they allow direct communication between the cytoplasm of adjacent plant cells, through the otherwise rigid cellulose cell walls.

How do vine roots take nutrients from the soil? One of the key concepts here is cation exchange. Roots are able to exchange hydrogen ions, which they pump out, for the cations attached to the negatively charged soil particles, such as clay and humus. Clay often carries a negative charge, whereas humus—decayed organic material—can carry both negative and positive charges, and so can hold both cations and anions. Cation exchange capacity (CEC) refers to the number of positive ions (such as calcium, magnesium, iron, and the nitrogen-containing ammonium ion) that the soils can hold.

When clay and humus have a negative electrical charge they are able to hold onto positively charged ions. Generally speaking, CEC correlates positively with soil fertility, because it determines how many plant nutrients the soils can hang on to. Soil pH also affects CEC: more acid soils (lower pH) have a lower CEC than more alkaline soils (high pH). One way to increase CEC is to increase the organic content of soils. This has the benefit of both increasing CEC, and thus fertility, and also increasing soil texture.

Without organic material or clay, soils find it hard to retain nutrients. For example, an excessively sandy or gravely soil will allow mineral ions to be rapidly leached from the soil by rainfall.

So, where do the mineral ions (nutrients) come from in the first place? It is not really from the bedrock, which would be the intuitive assumption on many. Some mineral ions might be bedrock-derived, but these would largely be in the subsoil. Low levels can come from rain, and some can come from the weathering of larger soil particles, such as stones and rocks. However, the bulk of soil nutrition will come from decaying organic material.

To get a better handle on this I spoke with Tim Carlisle who has studied soil science, but also has a deep understanding of wine from his current employment as a wine merchant. "We need to then look at the microbial activity in soil," he states. "This affects the speed and ability of soil to break down organic matter into mineral ions that can be used by the plant—and also aids the uptake of ions by plants. Because of this, no discussion about soil should exclude them, because whatever the terroir is, the level of microbial activity is an important and always overlooked element."

Carlisle points out that there are many factors that influence this microbial activity, but primarily water, food, and oxygen. "Oxygen is more available in a loose uncompacted soil," he says. "A soil that is overly compacted has little oxygen and so little microbial activity—the same is true of a waterlogged soil—which is one reason why porous bedrock and/or slopes are important, not just because of vine stress but because the microflora and fauna don't get drowned." The term microflora refers to the bacteria and fungi in the soil. Their existence also governs the level of microfauna, which refers to soil organisms ranging from single-celled protista to small arthropods and insects, through to nematodes and earthworms.

"The food they need is organic matter. If you visited a conventional agriculture wheat field you'd find there was very little organic matter in the soil, and, as a result, very little microbial activity. This is further diminished by crop spraying, hence the vicious cycle of needing to use tons of fertilizer."

During his studies of soil science, Carlisle looked at the effect of fungicides, herbicides, and insecticides on the soil microbes. He did this in two different ways. First, he took microbes from the soil, grew them in culture, and then studied the effects of dilute agrochemicals on their growth. "What I found in this was that fungicides over herbicides and insecticides kill off not just fungi (which includes yeasts and molds)," says Carlisle, "but also a high proportion of bacteria, and actinomycetes (I didn't do anything with algae), but also that herbicides and insecticides also killed off a proportion of all types of microbe, and restricted growth of others."

He also studied the overall microbial activity in the soil. "What this showed up was that untreated soil was healthier than anything with any kind of treatment, including one that was sprayed with fertilizer," reports Carlisle. "If you think about it, minerals are essentially the excretion of microbes. Too much excretion to soil will poison them and so spraying with fertilizer actually caused a check in microbial activity—it continued but at a lesser rate." He adds, "The thing that was by far the most interesting from a viticulture perspective was that one of the samples was sprayed with copper sulfate, which is permitted in organic viticulture. This sample was the one in which microbial activity was reduced by the most."

In 2010, Elda Vitanovi and colleagues published a study looking at the level of copper in vineyard soils in Croatia, which routinely receive about 3–5kg per (6.5–11lb) hectare of copper fungicides each year. They found that copper levels were significantly higher in vineyard soils than in other soils they looked at, and that 17 out of 20 vineyards studied officially fell into the category of copper contamination. A French research group reports that copper levels in vineyard soils typically range from 100–1,500mg per kg (2.2lb). They reckon that across the world, some eight million hectares are affected by this copper pollution, with half of them in Europe and one-eighth of them in France. It is clearly an issue that needs addressing. "It's not a problem for the quality of wine directly," says Claire Chenu, "because copper is little assimilated by vines. But it does lead to a decrease in the diversity and abundance of soil microorganisms, and a decrease in diversity and abundance of soil fauna (including earthworms)." Correspondingly, says Chenu, there will be a decrease in the soil properties and functions that these organisms help to promote, such as aggregate stability.

Franz Weninger, a biodynamic winegrower in Austria and Hungary, points out that some of the elevated copper levels in vineyard soils date back to the time when up to 30kg (66lb) of copper per hectare was applied annually. The current situation in Austria is that the maximum amount allowed is 4kg (9lb) per hectare and organic/biodynamic winegrowers can use only 3kg (6.6lb) per hectare. "We manage with tunnel sprayers to come down to 1–1.5kg [2–3lb] per hectare," he states. "This is about 0.15g per square meter [.005oz per square yard]." Weninger points out that while copper has a very toxic effect on soil life in the lab, copper in the field is about six times less aggressive to soil life. "I hope that in the long term organic and biodynamic winemakers can get rid of copper," he says, "but, believe, me glyphosate and the systemic fungicides are far more of a problem for the soil life."

RHIZOSPHERE INTERACTIONS

Out of sight and hidden underground, the root environment is relatively neglected. But this space, known as the rhizosphere, is home to some complex physical and biological interactions among the roots and soil-dwelling microorganisms.

The roots have been described as "rhizosphere ambassadors", facilitating communication between the plant and other organisms in the soil.

Plants secrete a wide range of compounds into the soil from their roots. These include amino acids, organic acids, vitamins, nucleosides, ions, gases, and proteins. Such secretions are important in maintaining the rhizosphere biology. This is illustrated by an interesting experiment in a Swedish forest, in which the photosynthates produced in the leaves were prevented from being transported to the roots by the stripping away of a ring of bark all around the diameter of the trunk. This removes the phloem layer of vascular cells, just under the bark, and would normally kill the tree. It resulted in a massively reduced amount of life in the soil within just a few days. The trees were clearly supplying the microbes and other soil organisms with nutrients.

Why do plants take part in this energetically costly charity work? It is because, eventually, it benefits them. Some of the chemicals released help with nutrient uptake. This can be an indirect effect, through reducing the soil pH (which can help with making some nutrients more available) or promoting the life of soil microbes that take part in nutrient recycling. Some root exudates can act as "chelators" of metal ions that are needed as nutrients. For example, iron is frequently abundant in the soil but is often present in a form that is hard for plants to access. Various root extracts can chelate (form chemical complexes) with iron that is then made available for uptake by the roots. Phosphorus is also often present in an unavailable form, but organic acids released by the roots can help make it more available.

Root secretions can also play a role in signaling: there's a chemical conversation going on between plants and other plants, and plants and microbes, for example. This can be both positive and negative. For example, isoflavone secretion by soybean roots attracts both a mutualistic fungus (good) and a fungal pathogen (bad). There's a phenomenon known as allelopathy, which is when one plant secretes compounds that discourage the growth of other plants nearby. Sometimes they even discourage the growth of the same species nearby (autotoxicity). Plants under attack by a pathogen can signal through their roots to their neighbors of the same species and thereby induce resistance in the plants that have not yet been attacked. This induced herbivore resistance can be direct (turning on defense strategies), or indirect (sending out volatile chemical signals that attract predators of the insect pest, as a type of airborne SOS signal). An example of this would be in the lima bean, where attack by spider mites causes root exudates from the plant under attack that prompts neighboring plants to release a volatile signal that attracts predatory mites that feed on spider mites.

One well-known example of a beneficial rhizosphere interaction is that of the mycorrhizal association. Some eighty percent of terrestrial plants have these mutually beneficial symbiotic relationships in which certain fungi become associated intimately with the roots and help with nutrient uptake (through massively increasing the surface area of root to soil contact), in exchange for lipids and carbohydrates from the plants. In legumes, a very special relationship occurs in which signals from the roots begin a series of interactions with nitrogen-fixing bacteria, which then cause the root morphology to change so that these bacteria become enclosed in specialized structures called nodules. The plant root secretions cause the bacteria to arrive and release chemicals that then modify the way the root cells grow. It's a brilliant arrangement. They get a home and nutrients while the plant gets a nitrogen supply.

The importance of rhizosphere interactions in grapevine growth hasn't been studied in detail, but the takeaway here is that what happens underground is complex and important. The biology occurring in the soils has been a neglected area of viticulture research, and there is more happening here than we are aware of. How we manage vineyard soils matters. There could be unintended consequences to the use of fungicides and herbicides. As discussed earlier, adding copper to the soil clearly isn't a great idea, but copper-containing fungicides are still widely used in organic and biodynamic viticulture.

ROOTS AND HORMONE SIGNALING

Vine roots are important to grape quality in a number of ways. Significantly, root growth causes hormonal signals to be sent to the portion of the plant above ground, and these act as instructions to tell the vine to modify its growth. Perhaps the best summary of what is happening here comes from the Western Australian plant biologist John Gladstones, in his book *Wine, terroir and climate change* (Wakefield Press, 2011), in which he brings together the existing literature and adds some theories of his own.

Gladstones points out that gibberellins (one of the major plant hormones) promote shoot internode growth through cell extension (the nodes are the parts of the stem between two buds), and these gibberellins are formed in the region of cell division behind root tips. They also promote the formation of tendrils rather than fruit clusters in the newly forming lateral buds. Presumably, if conditions are good for root growth, the vine is likely to favor vegetative growth, and gibberellins are sending up these instructions from the roots.

Cytokinins (another major plant hormone), produced in the root tips, promote new node and leaf formation, the branching of shoots, the development of existing fruit clusters, and the fruitfulness of newly forming lateral buds. Gladstones observes that warm spring soils promote cytokinin dominance; cool soils gibberellins. Warm spring soils are therefore a good thing for grape production.

Abscisic acid (ABA) is probably the most interesting of the plant hormones for wine quality, though. When roots signal to the vine that water stress is coming using ABA, the leaves respond by closing their gas exchange pores. ABA also acts in tension with another group of plant hormones, auxins, in controlling ripening. There is a kind of tug of war going on here. The auxins, produced by developing seed, slow down berry development, prolonging the pre-*veraison* phase while ABA pulls for berries to develop faster. In conditions of water stress, ABA signals to the berries to hasten development. A massive transfer of ABA to fruit coincides with *veraison,* possibly because of declining berry auxins at this stage. Therefore, root moisture stress is contributing to berry ripeness. Gladstones concludes that ABA is the primary hormone imported into the grape clusters that both triggers and continues to stimulate ripening.

A study by Dr. Hendrik Poorter and colleagues in Germany used magnetic resonance imaging (MRI) to look at root growth in potted plants. They were interested in finding out how big the pots have to be for experimental work, studying a wide range of different pot-grown plants.

The results emphasized how important root signaling is for the growth of the portion of the plant above ground. The pot size restricted growth for a wide range of species and doubling the size of the pot increased growth by 43 percent. The MRI results indicated that the plants were using their roots to "sense" the size of the pot, and signaling this to the rest of the plant.

Gladstones' assessment of the literature on vine physiology agrees with this idea. By means of hormones, roots are signaling to the portion of the plant above ground. The root structure is determined by soil conditions: deep soils with ample water supply prolong the phase of root development and signal the above-ground plant to keep growing; shallower soils with limited water or a texture that obstructs root growth reduces vegetative vigor. Low vigor is usually best for wine quality, especially for red wines.

WATER RELATIONS AND WINE QUALITY

One of the most important properties of vineyard soils is how they control water supply to the vine. Vine roots are designed to be particularly effective in taking up water and nutrients from the soil,

because, owing to their climbing habit in the wild, they are establishing themselves in soils already colonized by other plants. Supply vines with too much water and they will grow big, lush canopies and not put much effort into grape production.

Researcher Gérard Seguin carried out a famous study on the soils of the Médoc (Bordeaux) in the 1980s. He found that many of the best vineyard sites had poor levels of soil nutrients, but that this was compensated for by deep root systems. These top sites were frequently acidic gravels, and showed magnesium deficiency due to high potassium levels, as well as low levels of nitrogen. But aside from this, it was hard to correlate potential wine quality and soil nutrient levels. Seguin stated that, "it is impossible to establish any correlation between quality of wine and the soil content of any nutritive element." He adds that if there were to be such a correlation, you could give yourself a good chance of making great wine simply with the assistance of chemical additives to the soil.

Seguin's major conclusion was that the vital way in which the soils affected grape quality was through regulating water availability during the vegetative cycle of the vine. Moderate water deficit has been shown to reduce shoot growth (vigor), berry weight, and yield, and increases berry anthocyanin and tannin content—ideal for high-quality red wine production. Vine-water status is dependent on soil and climate characteristics, and soil influences vine-water status through its water-holding capacity. Seguin showed that in the Bordeaux vineyards, which are not irrigated, berry size is decreased and total phenolics are increased when vines face water deficits, resulting in higher grape-quality potential but lower yields.

A more recent study by Bordeaux-based terroir researcher Kees van Leeuwin, who has worked with Seguin in the past, examined this in more detail. Van Leeuwin and colleagues looked at 32 vintages in Bordeaux from 1974–2005 and found a correlation between vine-water deficit stress index and vintage-quality ratings. "The quality of red Bordeaux wine can be better correlated to the dryness of the vintage than to the sum of active temperatures," he concludes. In none of these vintages included in the study did the quality suffer because of excessive water deficit. Vintage-quality ratings don't correlate well with the average growing season temperatures, which is surprising. However, the story isn't a simple one: some vintages, such as 1982, were excellent with no real water deficit.

In Bordeaux, the vineyards that rarely experience deficit are either planted with Merlot, or with white varieties. Closer planting, growing higher canopies, and using rootstocks that only partially use soil-water reserves (such as Riparia Gloire de Montpellier) are vineyard interventions that can be used on sites that usually lack natural water deficit.

But water deficit is not always associated with higher wine quality. As part of his Ph.D studies, Jean-Sebastien Reynaud looked at the effect of soil water-holding capacity on wine quality in 23 different Vaud vineyards, in Switzerland. These were planted with Gamaret, a Swiss variety that's a cross between Gamay and Reichensteiner and is popular in Switzerland because of its resistance to botrytis. Many studies of water deficit have involved irrigated vineyards, but Reynaud looked at unirrigated sites over three vintages, 2007–2009. Water stress can have both positive and negative effects on vines. In drying soils, plant roots synthesize abscisic acid, which signals to the portion of the plant above ground and encourages grape ripening, and the partitioning of carbon resources to the fruit rather than the canopy. But it also causes the pores in the leaf (known as the stomata) to close, reducing carbon assimilation. If the stress is severe then leaves are lost. So, stress tends to increase soluble solids in the grape (generally good for wine quality, especially in reds) until it reaches a certain point and soluble solids decrease when the stress is too much. Reynaud found that water deficit improved wine color, but

there was no clear relation between water stress and the sensory attributes of the experimental wines he made. "In the Vaud conditions, vine-water stress was not the major parameter responsible for differences in wine quality," he concluded.

Perhaps Seguin's viewpoint that water-holding capacity is the key soil factor for wine quality—widely accepted around the wine industry, particularly for red wines—has overshadowed the potential importance of soil chemistry. The soils he studied were of a certain type, what Chilean terroir expert Pedro Parra describes as "geomorphic" soils (those where the soil isn't formed directly from the underlying bedrock). Parra suggests that "geological" soils, formed from weathered bedrock, may be quite different in this respect than geomorphic soils. Examples of geomorphic soils would include alluvial soils, such as those of Bordeaux, or wind-blown loess. In contrast, the soils of Burgundy and the Rhône would be geological.

THE IMPORTANCE OF CLAY

Earlier I touched on the different sorts of clays, and the implications of these for viticulture. Clays are made of the smallest of all soil particles, and because of their structure present an extremely large surface to volume ratio. They can hold water and nutrient ions very effectively. "I'd always thought that clay was an unlikely viticultural soil," says John Atkinson MW, who has a vineyard in the UK and is the author of an interesting research paper on terroir in Burgundy's Côte de Nuits. "It just seemed so charged with minerals and water, and therefore overly invigorating in temperate climates. I was therefore surprised to read in Denis Dubordieu's two-volume work on oenology that Petrus' smectite clay soil hydrically stressed the property's Merlot vines."

"I read around the subject," continues Atkinson, "and came across a report on Geelong's [Victoria, Australia] soils. The article makes the important distinctions between soil water capacity, availability, and extractability." Atkinson points out that clay soils hold onto their water, and their density makes rooting difficult. "Clays might appear humid, but extractability can limit availability," he adds. "This tied in with a paper I'd read by Kees van Leeuwen, which reported hydric stress occurring more rapidly in vines grown on a clay-based Bordeaux soil than they did on a more typical Médoc gravel soil."

Apparently, Petrus, the famous Pomerol estate, has more smectite clay than any other estate in Pomerol. "Smectite is a volcanic mineral that increases the internal surface areas of clays, and exaggerates their shrink–swell properties," says Atkinson. "It is part of the montmorillonite group of clays. Expansion of the clays closes the pores in the soil and makes conditions too anaerobic for root growth while impeding existing root function. Consequently, they impose a limit on the extractability of water by limiting root development. Conversely, as smectite/montmorillonite clays dry out, they shrink and crack, allowing rootlets to populate the developing capillaries. By contrast, Kaolinite and illite clays expand and contract very little."

"The relevance of all this is that one could model a viticultural regime based upon montmorillonite clay, warm summers, and irregular rainfall in which vines are nearly always under stress; even heavy rainfall wouldn't penetrate the soil, because the clays expand and seal at their surfaces. Petrus would be the paradigm of this sort of interaction."

The main focus of Atkinson's work on terroir has been in Burgundy. The famed vineyards of the Côte d'Or are on the side of a rift valley and, because of this, soil can vary over the course of just a few yards. Atkinson says that the best vineyards are those where there is limestone bedrock and plenty of active calcium carbonate, which helps create an open soil structure. The flocculated clays in these soils have the physical drainage properties of sand but can hold nutrients. The porous limestone soils can help with drainage because of their effect on the structure of the soil but can also act as a water reserve. The clay content in the soil is able to extract the water from the limestone reserve and make it available in limited quantities through the growing season. The red *grand crus* typically have smectite (swelling clays), while those with kaolinite clays are better suited to white wines.

Atkinson cites the work of Frank Wittendal, who used a statistical technique called principal components analysis (PCA) to look in detail at 2,816 specific *climats* in Burgundy. The *climat* is the Burgundian terroir unit, and represents a single patch of presumably homogeneous terroir. This may be a vineyard in its entirety or may just be a part of a larger vineyard. For example, Clos de Vougeot is a 50-hectare *grand cru* vineyard with 16 *climats*, while Corton, another *grand cru* vineyard, has 24 *climats*.

Wittendal described each *climat* in terms of 14 soil-description variables and four landscape/climatic variables, which were then fed into the PCA. The great significance of this work is that by using statistics, he was able to show that factors that would be assumed to be important—such

Above Limestone is highly regarded as a vineyard soil component. Bell Hill, in North Canterbury, New Zealand, was established because of the almost pure limestone subsoil, shown in this cutaway.

as altitude, aspect, parent rock, and gradient—weren't significant in separating out the different hierarchical levels in Burgundy's vineyards.

The PCA work shows that in terms of vineyard classification, the soil properties are really what matter, and these have precedence over altitude and slope. There is some evidence that east-facing vineyards are favored, but this is because facing east correlates well with interesting soil types, rather than the angle of the sun's illumination.

Wittendal's analysis was able to split the different *climats* up into three quite separate groups. Group one consists of colluvial soils. These are formed by the accumulation of fallen, eroded soils, which are retained on shallow slopes. Colluvium is made of fragmented rocks. Group two is noncolluvial compact limestone. This is able to retain water and possibly sequester it from deeper sources. Group three is alluvial soils, at the lower end of the altitude and slope indicators. There is just one *grand cru climat* with an alluvial soil, a Bâtard-Montrachet. Root growth in limestone and colluvial soils is quite different. In the limestone soils, roots go mainly down, searching for water. In colluvial soils, made of fragmented rocks, roots travel in all directions.

If just the red *grand cru* vineyards are included, then eighty to ninety percent of the clayey limestone soils with active carbonate are eliminated, and the majority soil type is colluvium. That is, the *grand cru* vineyards had a high proportion of gravely, colluvial hillwash in their subsoils. Atkinson suggests that the significance of this is that the best vineyards put the vines into the optimal deficit zone, and it's the hydrology properties of the vineyard that determine its quality potential. This is similar to the findings of Seguin and others in Bordeaux, where the alluvial sand and gravel soils, coupled with a retreating water table in the summer, puts the vines into deficit. In the best Bordeaux vineyards clay lenses (a lens-shaped block of clay) run through the soil helping to buffer water availability, keeping the vines supplied with just enough water but letting them experience mild deficit.

MINERALITY IN WINE: CAN WE TASTE THE SOIL?

This brings us onto the interesting question of minerality. "Minerality is the perception of the rocks in the soil, by the palate," claims soil scientist Lydia Bourguignon who, with her husband, Claude, forms one of the most respected and influential consultancies on the role of vineyard soils. "We hold geosensorial tastings," she continues. "It is a sort of taste training in which you actually touch, even taste, different rocks to then be able to find the same sensations on the palate with the wine. For example, touching granite gives a cold impression, while limestone seems warm."

What is minerality as applied to wine? Can we define it as a tasting term? Do we even mean the same thing when we each talk about "mineral" wines? And can we try to link minerality to chemicals present in the wine? These are all complicated questions to answer, and for many people the whole concept of minerality is so muddled and confusing they would rather avoid it all together.

"Minerality" is a really useful descriptor; many of us use it frequently in our tasting notes. Yet it's also a term that means different things to different people. I know what I mean when I encounter some characteristic in a wine that makes me think "mineral," but I can't be sure that when others use the term they are referring to the same thing. Not only that, minerality may well be a sort of syndrome, like some medical conditions, when different underlying factors cause symptoms that look quite similar. I also suspect that it is sometimes used as a way of praising a deliciously complex wine, in the same way that "long" is often thrown in to a tasting note when people really like a wine but have run out of more concrete descriptors.

HOW EXPERTS USE THE TERM

What do different tasters mean by "minerality"? "'Mineral' is interesting," says well-known English wine writer and contributing editor for *Decanter* magazine, Stephen Spurrier, "but it did not exist as a wine-tasting term until the mid-1980s. During most of my time in Paris I don't think I ever used the word," he recalls. "I think this was because most French vineyards were overproducing, chaptalizing, and doing all those things that mean minerality—which has to come from the soil and nothing else—was not looked for and not present." Spurrier does use the term quite a lot in his tasting notes. "I probably associate minerality with stoniness, but then stones are hard and minerality is generally 'lifted.' As a taste, it just comes into my mind and I very often find myself writing 'nice minerality on the finish.' I suppose it is easier to describe what it is not, that is, it is not fruit, nor acidity, nor tannins, nor oak, nor richness, nor fleshiness. It is not really a texture, either, for texture is in the middle of the palate and minerality is at the end. I think it is just there, a sort of lifted and lively stoniness that brings a sense of grip and also a sense of depth, but it is neither grippy (which is tannin), nor deep (which is fruit)." Spurrier adds, "No wonder we are all a little confused."

I asked another well-known wine writer, Jancis Robinson, about her use of "minerality" in tasting notes. "I am very wary of using it because I know how sloppily it has been applied," she replied. "In general, I try to use it as little as possible and be a bit more specific. 'Wet stones' is a favorite tasting note of mine but there is sometimes something 'slatey' about some Mosel wine and 'schistous' (grainier) about some St-Chinian and Catalan wines from both sides of the Pyrenees, I think. But when wine definitely doesn't smell of anything fruity, vegetal, or animal, I might use it."

Noted French critic Michel Bettane describes minerality as "a fashionable word never employed in the 1970s and 1980s," agreeing here with Spurrier that it is a fairly recent invention as a tasting-note term. Bettane says that for many tasters it is a "politically correct" term, describing a "wine nonmanipulated by the winemaker and from organic viticulture, aromas and fruit being the sign of manipulation or the lack of expression of origin." He continues, "For me the only no-nonsense use is to describe a wine marked by salty and mineral undertones balancing (and not hiding) the fruit, more often a white wine rich in calcium and magnesium as many mineral waters are. For a red wine I have no idea, with the exception of some metallic undertones (iron in Château Latour or copper in Nuits-Saint-Georges les Pruliers)."

Jordi Ballester, a researcher from the Centre des Sciences du Goût in Dijon, in Burgundy, has been studying the use of the term minerality by lexical analysis. He has noted that it's a term being increasingly used, but without a clear definition. He and his colleagues compared the ratings of 34 wine experts and a trained panel, looking at the sorts of tastes and smells that people describe as "mineral." They found widespread differences among the tasters, and while there was some agreement about definitions when people were quizzed, the use of the term differed in practice. Interestingly, though, some of the subgroups used the term in similar ways, suggesting that there is a cultural basis for its use. The Sancerre winegrowers, for example, agreed on what they thought minerality was. So maybe minerality is in part a local concept?

TAKING MINERALITY LITERALLY

It is interesting that Bettane brings up mineral water. Frequently, you will see on the label of a bottle of mineral water a listing of the different levels of mineral ions present, which depend on the source of the water. Different mineral waters do tend to have subtly different flavors if they are compared side by side. Presumably, these differences in flavor, subtle as they may be, all come down to the mineral composition, although there are plenty who will contend that most mineral ions don't taste of anything.

This is the first definition of minerality we are going to explore: the literal one. In this definition, minerality in a wine context is a result of mineral ions, present in the soil, which find their way into grapes, and then affect the flavor of the wine.

If a soil has mineral ions in it, the roots will take these up passively, along with the water that the roots sequester. However, roots are also able to take up mineral ions selectively in certain circumstances. Some people object that the typical differences in mineral-ion concentration, such as those found in different mineral waters, would not be noticeable against the backdrop of the other flavors present in wine. However, the levels present in wine seem to be considerably higher.

"Minerals can be detected while tasting a wine," says Olivier Humbrecht, one of Alsace's leading winegrowers, famous for his advocacy of biodynamic viticulture. "It is the fraction on the palate that makes the wine taste more saline or salty. High acids or high tannins do not mean that the wine has lots of minerality. High salt contents make the acidity more 'savory' and therefore less aggressive. Good minerality makes one salivate and want to have another sip or glass or bottle."

Gerd Stepp, a consultant winemaker, has some interesting perspectives of minerality. He broadly falls into the literalist camp. "For me, there are two forms of minerality that influence a wine's qualities and characteristics," he explains. "Firstly, perhaps most obviously, the wine's mineral content, which is about taste and texture when tasting a wine. It's much like when drinking mineral water of a high mineral/salt content, there is a flavor/taste and an almost 'osmotic' experience, perhaps similar to drinking seawater, just much less concentrated and less salty." Stepp thinks that mineral waters taste differently, and can sometimes even seem a little salty, depending on the source. Stepp also notes that, according to his reference books, the mineral content of wines fluctuates between 1.5–4g per liter (.05–.14 oz per 34 fl oz). "It seems the soil's exchange capacity of ions correlates with the mineral concentration of a wine," he states. "Also, a cold-stabilized wine has lower potassium content than the same wine not stabilized, and it tastes different and has less flavor, perhaps even less complexity." But he points out that there is a problem with this theory, in that the stabilized wine will still seem to have the same qualitative flavors. "If from slate it still tastes like a wine from slate, just less intense," he says. "There is a matrix factor, as well, such as that residual sugar, acidity/pH, carbon dioxide, sulfur dioxide, malolactic fermentation, and oak, for example, can enhance or reduce what we taste as minerality."

For Stepp, there is also a second form of minerality. "I am certain there is an influence on the wine's flavor characteristics through the geology/soils where the grapes are grown, which correlates to the terroir's unique minerality," he claims. "I understand researchers trying to prove that vines don't actually take up minerals from the soils and that these minerals aren't in the finished product. But how can it be explained that wines made from the same grape variety, vintage, and region have such different qualities depending on the soils? It must have an influence, detectable or not. Also, where would those 1.5–4g per litre of minerals come from? But I don't believe that we just taste the actual minerals from the soils, it is definitely much more complex."

The anecdotal evidence suggesting that some terroirs create wines with much more "mineral" character is pretty strong. Soils matter. As an example, I was recently in Hungary, discussing the link between flavor and soils with a winegrower on Lake Balaton. He pointed out that a distinct soil type on part of his farm produced unusually long-lived white wines, a property he ascribed to the mineral characters in the wine conferred by the soil. He made a comparison with the same variety grown by a neighbor on different soils. His observation was that while this neighbor was a talented winemaker, the neighbor's wines, made from the same variety, never lived as long as his own. And the long-lived wine did have a distinctly mineral character to it: a sort of textural character that was hard to describe, but which seemed, for want of a better description, to be "mineral".

There is a very interesting pair of scientific studies in support of this notion of minerality in wine. Back in 2000, a plant researcher from Germany, Andreas Peuke, grew Riesling vines in pots containing three different soils from Franconian vineyards: loess, Muschelkalk (seashell lime), and Keuper. He collected sap from the vines and analyzed its chemical composition and found differences among the different soil types. In Muschelkalk soil, carbon, nitrogen, and calcium were present in the greatest concentrations. Sulfur, boron, magnesium, sodium, and potassium were greatest in Keuper, and the concentrations in loess soil were intermediate. Aqueous extraction of the soils resulted in a two-fold greater concentration of total solutes in Keuper extract compared with Muschelkalk, and more than threefold than in loess.

A few years ago, Californian winegrower Randall Grahm carried out some interesting experiments. In his quest to try to understand minerality better, he actually put some rocks into tanks of wine. However, in this case the rocks were in a wine environment at low pH and are therefore likely to release more "minerals" than if they had been in the ground. The rocks also had the side effect of raising the pH of the wine, which can change the flavor. "Our experiments were incredibly simplistic and gross in comparison to the very subtle chemistry that occurs in mineral extraction in real soils," Grahm recalls. "We simply took interesting rocks, washed them very well, smashed them up and immersed them in a barrel of wine for a certain period, until we felt that the wine had extracted some interesting flavors and we were able to discern significant differences between the various types. We initially screened a number of different rocks with bench trials and ultimately decided on a few for larger scale experimentation: granite, Noyo [River] cobblestone, black slate, and pami pebbles," he recalls. Grahm saw major changes in the texture and mouthfeel of the wine, as well as dramatic differences in aromatics, length, and persistence of flavor. "In every case, low doses of minerals added far more complexity and greater persistence on the palate. It is my personal belief that wines richer in minerals just present way differently." He adds that, "they seem to have a certain sort of nucleus or density around their center; they are gathered, focused, cohered the way a laser coheres light. It is a different kind of density relative to tannic density, somehow deeper in the wine than the tannins."

REDUCTION AS MINERALITY

Some people use "mineral" to describe aromas of wine; it's something they get on the nose. In this case, it could be that tasters are ascribing minerality to what is in reality the presence of certain volatile sulfur compounds in the wine, also known by the term "reduction." In its most raw state, reduction is caused by hydrogen sulfide, and smells like rotten eggs and sewers. This is rare in a finished wine and would not be classed as mineral.

Far more common is the presence of complex sulfides and mercaptans (also known as thiols). These sulfur compounds, like hydrogen sulfide, are produced largely by yeasts during fermentation. Their expression depends on their concentration and the context of the wine, but in some cases they can give a flinty or struck-match aroma that can be quite "mineral."

There is good reason to suggest that flintiness in white wines is a result of some low-level reduction. Great white Burgundies frequently show a little of this good reduction: a matchstick element to the nose is complexing. Some New World Chardonnay producers are now beginning to work out how to achieve this through winemaking practices. There's also a link here with terroir: some sites naturally have nutrient deficiencies that can stress the yeasts a little and cause them to produce more of these volatile sulfur compounds.

But scientists tend to prefer this second definition of minerality because they dispute the first—the more literal definition in which minerals in the soil end up flavoring the wine. This tends to make some believers in minerality act a little defensively when discussing the subject. "I fully understand that when I use the term it may have no scientific validity," says Jasper Morris, wine merchant and Burgundy expert. "I have been told by enough geologists that you shouldn't call Chablis flinty, because flint is not soluble in water and therefore it can't have a taste. But it is an image." Morris illustrates his point by comparing two Burgundy vintages. "I would use two vintages of white Burgundy to illustrate this point: 2007 and 2008. The acidity is higher in 2008, and when I taste those wines I feel that they are acid and not especially mineral. In 2007, the acidity is a little lower, but I find the wines distinctly mineral. By that, I mean that they have a fresh zingy zest to them that in my mind puts up the single word 'mineral'. In 2006 the wines are fat, rich, and round, and some of them have enough acidity to provide balance, but you don't feel mineral when you taste the wines." He continues, "Also in Burgundy, in terms of the hillsides, you expect minerality in those soils, which have more active limestone, and you expect less in those soils which are more clay based. For example, with any vineyard in any village that includes the words 'charmes' in the name, you rarely get the concept of minerality." Morris is frustrated that the scientists seem to be criticizing minerality, without offering an alternative explanation: "Scientific fundamentalists are denying that we get it right rather than offering any certainty in the other direction."

Paul Draper of Ridge Vineyards, in California, is a winegrower who believes in minerality from the soil, even in the face of questioning from scientists. "Though I am well aware of what soil scientists say about minerals or other elements in the soil and the impossibility of them traveling through the vine and into the wine, the roots deep enough into those minerals are affected and the wine shows that effect," he states. "I think of minerality as a wet-stone quality in a wine. Our subsoils at Monte Bello are limestone and at times are at the surface or a meter below. In other places our backhoe pits find them several meters down. Perhaps seventy percent of our vine roots are deep in the limestone. I have seen minerality in some shales as well, so I don't think the effect is necessarily limited to limestone. We see the most marked minerality (crushed rock, perhaps flint are other descriptors) is in our more eroded blocks where the limestone is closer to the surface. In the youngest blocks where pits have shown considerable limestone we don't see the minerality as yet but expect to when the roots are deeper."

Minerality remains enigmatic. As we begin to understand more about it, the picture seems multifactorial, with different mechanistic underpinnings for what wine tasters describe as mineral in their tasting notes. From a personal viewpoint, I used to favor the more established scientific viewpoint, assuming that volatile sulfur compounds could explain much of minerality. But I'm increasingly drawn to the idea that minerals in wine, derived from the soil, could be affecting wine flavor in interesting ways and, in particular, helping to create long-lived compelling white wines.

SPECIFIC DEFICIENCIES OF MINERALS

Famed vineyard soil scientist Claude Bourguignon points out that all the compounds produced in a living system are produced by enzymes, and many of these enzymes have metallic cofactors, such as magnesium in chlorophyll, or manganese and magnesium cofactors in enzymes that build up monoterpene molecules. Micronutrients such as these metallic cofactors all come from the soil. Bourguignon points out that we still don't know all the enzyme pathways involved in the synthesis of aroma compounds or their precursors. His assertion is that hydroponically produced fruits have no taste, because without soil microbes there are no micronutrients and thus no aroma, because enzymes need these micronutrients as cofactors.

I suspect that Bourguignon may be onto something here, but in a slightly different way. Rather than great terroirs offering a surfeit of micronutrients, it may be that some offer slight but meaningful deficiencies, such that vine growth is not overly impeded, but the deficiency either causes a reduced enzyme activity in producing specific flavor compounds (or precursors), or results in musts that are slightly deficient in certain yeast nutrients, resulting in complexing characters in the wine—perhaps through a little reduction early on that then resolves into interesting complexity in the final wine. While plants wouldn't get far without chlorophyll, some of the flavor-active compounds or flavor precursors are nonessential secondary metabolites, so it is expected that there might be some variation in their levels in grape berries. This is all speculation, however. It would be great to have some more decent scientific studies on this topic.

THE TASTE OF TERROIR

"The vine may be cultivated advantageously in a great variety of soils... The sandy soil will, in general, produce a delicate wine, the calcareous soil a spiritous wine, the decomposed granite a brisk wine."
James Busby, *A Treatise on the Culture of the Vine, and the Art of Making Wine*, 1825

One of the frustrations of the topic of terroir is that while many people have described the physical characters of vineyard sites, there have been very few serious attempts to link soils with wine flavor. Many of us in the wine trade who

taste regularly will have anecdotal experience of the way that wines from different soils taste different. For example, there's a dramatic difference between Touriga Nacional produced from the schist of the Douro Valley in Portugal and wines from the same variety grown in granite soils (such as the granite at Vale Meão in the Douro Superior, or with a cooler climate, the sandy granitic soils of the Dão region).

A similar comparison can be made in the Northern Rhône where Syrah is grown on both schist and granite. The granite soils produce lighter-bodied wines with more freshness and aromatic purity, tending to more floral notes. Wines from schist tend to be richer and more structured. I'm sure that Mosel experts could probably pick blind Riesling wines made from the different sorts of slate that form the majority of the vineyard soils here.

"If you taste the wines of Europe and start looking at the soils, these patterns start to emerge, so that one kind of soil will give you a consistent stylistic imprint on the wine," says Mike Weersing, who founded Pyramid Valley Vineyards in New Zealand's North Canterbury region. He cites the differences between clay and limestone soils. Clay soils give flesh to Pinot Noir while limestone gives structure. Limestone alone can often be a little too thin, lean, and mean, so he's looking for both together in the topsoil. "What do clay and limestone give Pinot?" asks Weersing. "It gives it flesh, and it gives it the girdling of the flesh, which gives it structure and length and energy. You can see this in Burgundy. If you look at the typical Burgundian slope, you have hard limestone at the top, then soft limestone, clayey limestone, and then clay. And on the same slope, if you compare the wines, the wine from the hard limestone and soft limestone at the top (where there is not much clay) is very tight and vibrant but does not have much roundness or richness. Once you move down into what the French call the kidney of the slope, where the clay and limestone mix, you get depth and richness, but you also have structure. Then when you move off this into the clay Bourgogne soils you can have wines that are pleasantly fat, but they don't age well."

"The French call this clay-lime complex 'argilo-calcaire'," says Weersing, "and their simple observation (made over something like 1,200 years) is that Pinot Noir and Chardonnay, to be at their best, require the mix—neither only clay nor only lime at the surface, but a marriage of the two." He says that the bedrock isn't so important, although fractured limestone or chalk are perfect, because they release excess moisture, but retain sufficient water for the vine to drink evenly, without irrigation, throughout the growing season.

Terroir expert Pedro Parra thinks it's important to understand soils because this can inform winemaking decisions. "What is important is to try to understand in which area we are located, if we want to relate terroir to wine," says Parra. "If we are on volcanic soils and we don't work well, then we end up with bitter tannins," he suggests, referring to red wines. "It is easy to get bitter tannins on volcanic soils. If you are on granite you risk going to dry tannins. If you are on schist, depending on the type of schist, you can go to dry tannins or bitter tannins. On calcareous soils, the risk is green tannins."

Parra states: "When you identify the lithology [the parental rock] you can understand the winemaking you need to do to make good wines." But, he adds, it is not always so simple. "In Apalta [Chile, one of the areas he consults in], there is a mix of volcanic and granitic soils, so you could have dry or bitter tannins."

CONCLUDING REMARKS

It is clear that soils appear to be hugely important in shaping wine style. Climate determines whether you can grow a grape variety successfully in a particular location, but the soil type may determine wine style and quality, given a suitable climate. This is currently an area of great interest to winegrowers, but there's still a lack of good research linking soils to vine growth, and onward to berry composition and then further on, completing the loop with the implications for winemaking and the sensory characteristics of the resulting wine.

4 Climate

Agricultural crops are at the mercy of the weather. This is particularly true for vineyards, where we are all too familiar with the idea of vintage variation: the weather of the season imprints itself on the wine in often memorable ways. Vines are particularly sensitive to climate. There are thousands of grape varieties grown commercially, producing a bewildering array of different wine styles. But the performance of each variety is highly dependent on climatic factors, as well as the soil type. The sensitivity of grapevines to climate caused viticulturist Richard Smart to call wine "a canary in the coal mine" for climate change, because it is the world's vineyards that are feeling the effects of climate change ahead of many other crop species.

In addition to performing best within narrow climatic bands, grapevines are sensitive to weather at specific times in their growth cycle. Early on, they are at risk of spring frosts, which can devastate young growth. Warmer springs encourage vines to start their growth cycle earlier, which then leaves them susceptible to frosts— an increasing problem in many regions. Then there is flowering. Cold or wet weather during flowering can cause yield losses, with obvious economic impact. And warm, damp conditions during the growing season can encourage downy mildew, to which *Vitis vinifera* is highly susceptible, and then toward harvest time, wet conditions elevate the risk of botrytis bunch rot.

As harvest approaches, the developing grape berries are sensitive. Too little light and warmth, and they struggle to reach "sugar ripeness," the point at which sugars rise and levels of organic acids decrease. Too much light and warmth, and they reach sugar ripeness before hitting physiological, or "phenolic," ripeness. This means that polyphenolic compounds, such as tannins and anthocyanins in the fruit's skin, never undergo natural modifications to smooth out their astringency. Because grapes should be harvested only at phenolic ripeness, growers are then forced to correct problems in the winery.

This chapter sets out to explore the science behind climate and wine. I'll begin with an issue very much on the minds of winegrowers worldwide: climate chaos. And then I'll tackle ways of measuring and categorizing climate, using climatic indices.

CLIMATE CHAOS

The issue of global warming (or, more correctly, climate chaos) is a huge current concern for the wine industry. Carbon dioxide is one of a number of greenhouse gases that perform an important role in the atmosphere. They allow solar radiation to warm the planet, and then they act as a sort of gaseous insulating layer that stops some of this heat from escaping. Without the greenhouse effect, surface temperatures on Earth would be some 30°C (54°F) lower, and life as we know it wouldn't exist. But human activity has increased atmospheric carbon dioxide levels, largely as a result of the burning of fossil fuels. Consequently, the greenhouse effect has become more intense, and global average temperatures have risen.

The Intergovernmental Panel on Climate Change (IPCC) is the United Nations body for assessing the science related to climate change. It provides policy makers with objective information on this complex subject. Its latest report, from 2014 (the Fifth Assessment Report; see www.ipcc.ch), makes disturbing reading. It shows clear trends of increasing average temperatures, rising sea levels, and increased melting of the polar ice caps. These changes all occur in tandem with increasing atmospheric greenhouse gas levels. Whatever we do to reduce greenhouse gas emissions, there will still be future temperature rises because of the amount that we have already put into the atmosphere, but the lower we manage to get our emissions, the lower these rises will be.

WINE PRODUCTION IN A CHANGING CLIMATE

Back in 2005, the wine world was shaken by a study by climatologist Dr. Gregory Jones, then working at Southern Oregon University. Jones and his team pored over 50 years of average growing-season-temperature data from 27 wine regions and compared them with expert assessments of vintage quality. They also ran the Hadley Centre climate model to look at the projected temperature changes in these wine regions up until 2049. The trends that they identified suggested that the wine world was facing a very uncomfortable ride. Some 15 years later, it seems that these predictions are sadly true. And more change is to come.

Records in Burgundy stretching back to the medieval era show that warm periods similar to that of the 1990s occurred in the 1380s, 1420s, 1520s, and 1630s, each time followed by cooler periods. But as Jones and his team in Oregon found, the second half of the 20th century witnessed a unique trend, with an average rise of 1.26°C (2.3°F) in growing-season temperatures across the world's wine regions. And there has been no sign of cooling since, with 2003 the hottest year since these records began.

For some places, that trend has been a boon. For instance, English wine, once a curiosity, is now being taken seriously. In 2011, for example, there were 419 vineyards, 124 wineries, and 1,384 hectares of vines planted. As of 2019, there are now more than 3,000 hectares of vines in the ground, and 2018 was a record vintage, with some 15 million bottles produced.

Famed German winegrower Helmut Dönnhoff feels that the degree rise in average temperatures since 1988 has also been kind to German vignerons, saying that the "acidity is lower and the grapes are riper, so the quality of wines has improved." In the past, German Rieslings from cooler vintages often struggled to attain ripeness, and ended up with searing levels of acidity. His son, Cornelius, says that things might have gone too far in the other direction. "My grandfather fought for ripeness," he says. "Today I am fighting against ripeness. Now it isn't a problem to harvest healthy grapes with over 100 Oeschle." (Oeschle is the German scale for measuring ripeness and sugar content.) Dönnhoff says that now they need to manage photosynthesis in order to avoid too much sugar ripeness in the grapes, but at the same time he needs to protect the grapes from too much sun. "We have to bring drown the leaf-to-grape/fruit ratio," he says. "The middle part of the Nahe is a humid area, and we can't leave the grapes in the shade. We want to open the canopy but we also need to shade the grapes a bit. It is like a ride on a razor blade."

It is Jones' predictions for the coming decades that have winegrowers worried. These suggest an average increase in growing-season temperature of 2.04°C (3.6°F) between 2000 and 2049, with the largest predicted change in southern Portugal (2.85°C/5.1°F) and the lowest in South Africa (0.88°C/1.6°F).

For many, this could be disastrous. Water availability for irrigation is also a concern. Rising temperatures also bring an increased frequency of extreme weather events, and climate unpredictability. In some cases, the careful matching of grape variety to vineyard site may have to be reconsidered. This is not a simple solution: grapevines take at least three years to produce a crop, and only begin producing peak-quality grapes after 10 or 15 years. And, in most European regions, growers are allowed to plant only authorized varieties.

As evidence of how seriously changes in climate are being taken, in 2019, Bordeaux announced a change in the rules regarding permitted grape varieties. Seven new varieties were approved by the wine producers' syndicate for use with Bordeaux and Bordeaux Supérieur wines. These were the Portuguese variety Touriga Nacional, Iberian star white Albariño/Alvarinho, southern French cross Marselan (a mixture of Cabernet Sauvignon and Grenache), Petit Manseng, Castets, Arinarnoa, Liliorila, and Baroque. They will be allowed as ten percent of a blend.

THE WINE WORLD BEGINS TO TAKE CLIMATE CHANGE SERIOUSLY

Climate change is a major issue for the wine industry. Indeed, some are suggesting we should be referring to it as climate chaos. It's an issue that the wine world is beginning to address. I've recently attended three wine industry conferences on the topic, and another that included this as a main topic. The first was in Montréal in November 2017, at a meeting called "Tasting Climate Change" organized by Michelle Bouffard, a local wine writer and consultant. The second was a conference organized in Porto titled "Climate Change Leadership" in March 2019 (see page 60), and the third was the "Familia Torres Climate Change Course" held in Vilafranca de Penèdes, Spain, in April 2019 (see page 62). Finally, at the Sauvignon Blanc conference in Marlborough, New Zealand, in February 2019, climate was a major focus. The message from all the meetings was the same: action is needed now.

Steven Guilbeault, CEO of Équiterre, began the Montréal event by presenting an overview of the situation globally. Carbon dioxide concentrations are currently the highest they have been for 800,000 years, and this rise is through human activity. The trends are very clear:

our planet is getting warmer and warmer. So, what are the implications if the planet warms by 3°C (5.4°F) in the coming years? This is the projection by scientists of the sort of scale of change likely over the next century. This level of change would likely have dramatic impacts on the life on planet. Let's remember that when Canada was completely covered in ice, average temperatures were only 4°C (7.2°F) cooler. He pointed out that 1976 was the last time that average global temperatures were below average.

Risks include increasing drought, especially closer to the equator. But they also concern wine regions: in Napa, the severe droughts mean that if you now look at the fire-hazard severity map, there are lots of areas with a huge risk. Such severe droughts in California have resulted in legislation to cut water use by 35 percent. Overall, globally, there has been a severe increase in the cost of natural disasters. Fortunately, people have started paying attention to this problem. Guilbeault described how he recently attended the Paris Agreement, where there were three rooms meeting simultaneously, each with 5,000 people. The world is starting to recognize the severity of the situation.

How bad are things? Current policies take us +3.6°C (6.5°F) in the next 100 years. If pledges made in Paris are kept, then temperatures should rise by +2.7°C (4.9°F). It is not enough of a change but it is significant. All the modeling shows that if we pass the +2°C (3.6°F) mark, the climate will probably get out of hand, so this is a good first step but we need to do more.

What's happening? There have been investments in renewable energy production that, since 2010, have surpassed those in fossil fuel. And while fossil fuel is still the dominant source of energy, the prospects are not good for them (fortunately). As an example, the value of coal companies in the USA was $63 billion in 2011, but this had fallen to $4.7 billion in 2016. Even the Kentucky coal museum is switching to solar power (this is where the coal industry started in the USA). It might surprise many to hear that China is a solar power giant. They are responsible for 70.5 percent of the world market of solar energy, and this is just a decade after they began with it. They are shutting down coal plants and replacing them with renewables.

The second Montréal speaker, the aforementioned Dr. Gregory Jones, began by pointing out that the wine map is changing. China and Russia, for example, are planting more vineyards. There are also a lot of fringe cool-climate producers, in latitudes we would not have dreamed about a few decades ago. These changes are in response to growing demand and changing demographics, with new markets and styles of wines emerging. There are also new purchasing trends, such as vending machines. There are changes in the tastes of writers and critics. There is a large movement in the production of bulk wine. There is a keen interest in organic and biodynamic production. But most of all, underlying these changes is climate change.

One of the effects of climate change is that the dormant periods have got warmer in many regions, so the same cold hardiness doesn't develop, yet the winter freezes and extremes are still present. And warmer soil temperatures result in earlier budbreak, but the same frosts still occur. There are not dramatic changes in flowering, but cloud cover and rainfall can be more frequent affecting fruit set. Another observation is longer growing seasons and higher heat accumulation, but with high temperature variability. Also, lower diurnal temperature ranges occur during the final stages of ripening. Jones explained that on average phenology has shown a shift of between five–ten days, and the phases between these stages of growth have been compressed, too. There have been climate-related changes in soil fertility and erosion. Carbon dioxide levels are higher, but Jones says that there is not enough research to say anything about this and its impact on viticulture, although research is underway looking at how vines respond to higher levels of carbon dioxide in the air. Other issues are water availability and altered disease and pest susceptibility.

The +1°C (1.8°F) change over the 1950–2000 period has changed the viticultural map. For cool-climate wine production, some regions are now too warm whereas others have become suitable that were not before. Looking to the future, Jones used the example of the Willamette Valley in Oregon: +1°C (1.8°F) warming causes the climate envelope to move, and the capability for other varieties increases, while still allowing the region to grow Pinot Noir and Chardonnay. But a 2°C (3.6°F) rise pushes the envelope completely out of the spectrum of what can be grown today.

Continued warming of the world's wine regions is highly likely: the next century will see 1.5–4.5°C (2.7–8.1°F) warming. If this was a straightforward rise in average temperatures, then playing to the averages is easier to deal with. But the fact that there are more extremes makes things harder. If it was simply a question of rising average temperatures, then planting new varieties or moving to cooler areas would be an answer. But the variability we are likely to experience makes confronting change much trickier, with altered ripening profiles creating challenges in managing timing of acid, sugar, and flavor.

Jones outlined ways the wine world can adapt to changing climates. Wine grapes have a huge genetic pool of potential for adaptation, but this will mean breeding new varieties. There is also the landscape potential—the need to understand where to plant grapes and how to manage the vines. There are alterations to canopy geometry and even the use of shading materials. Scion/rootstock combinations need to be understood better, along with grapevine-water-use efficiency and irrigation management.

THE PORTO PROTOCOL

The Climate Change Leadership conference was held in Porto in March 2019. Headed up by Adrian Bridge of Taylor's Port, it brought together many players in the wine world to discuss specific actions that can be taken to mitigate climate change. It also saw the launch of the Porto Protocol. Wine companies can join by signing the declaration, but must commit to:

- Do more than they are doing at the moment;
- Guide their activity by good environmental practices and principles;
- Promote, in the respective value chain, the principles that characterize good environmental practices;
- Seek to identify opportunities for cooperation with other Porto Protocol members and with external entities regarding climate change;
- Execute projects oriented toward the sustainability of the organization;
- Disseminate good practices and case studies to The Porto Protocol platform;
- Participate in the initiatives promoted by Climate Change Leadership Porto, to support The Porto Protocol;
- Publicly report on their development performance;
- Appoint a delegate to manage the relationship with The Porto Protocol.

It lacks any certification or proper regulation, but as a way to encourage change, it is a start. As of August 2019, their website listed just over 150 members.

THE CARBON FOOTPRINT OF WINE

It is for this reason that the wine trade, along with the rest of society, has recently taken a keen interest in its carbon footprint. While the carbon footprint of the wine trade is just a small fraction of the total amount, we still, like all of society, have an obligation to challenge ourselves to reduce this footprint. But what exactly is the carbon footprint of wine, and what can be done about it?

There are two stages in the process of identifying the carbon footprint of wine. The first is to list all the processes, from grape to wine glass, where there is a net production of carbon dioxide (or other greenhouse gases) that otherwise would not have occurred. The second, a more difficult task, is to attempt to quantify these processes. In the vineyard, agricultural operations involving tractors or other vehicles are likely to be a big contributor. The fewer the passes through the vineyard by tractor, the lower the footprint. Some wineries use biodiesel to power their tractors. Alternatively, some wineries are considering the use of electric tractors that are charged by solar power, which may be an attractive option in the future.

Sprays have two levels of footprint: first, the energy required to manufacture the spray, and second, the fuel burned in delivering the spray. This calculation is complex: while a practice such as manual tillage of the soil may seem more environmentally friendly than the use of herbicides, it may require more mechanical input and thus have a higher carbon footprint. Likewise, biodynamic preparations have a carbon footprint in preparation (they are not usually dynamized by hand) and in application to the vines and the soil.

Pumping water for irrigation uses power and therefore has a footprint, as may the wastewater treatment process that is part of being a good environmental steward. Frost protection involving the use of propellers, burners, or helicopters is likely to be a rare event but involves some level of footprint.

Cooling the winery's buildings (or heating them in the winter in cooler regions) may require quite a lot of power. The more efficient the winery's design and the better the insulation, the less energy will be needed. Practices such as opening winery doors or windows at night to allow cooler air in may help passive cooling of the facility. Pumps in the winery use power, as does lighting the winery where this is required. Switching to modern energy-efficient lighting will help, and in some regions subsidies may be available.

Cooling tanks need major energy input. Many fermentations take place at low temperatures, and at various stages in winemaking, significant cooling of large volumes of wine is necessary, for example during cold stabilization. Techniques being developed to carry out stabilization of white wines without chilling to very low temperatures could result in reduction of winery footprint. But products such as metatartaric acid and carboxymethyl cellulose (CMC) have a manufacturing footprint that needs to be considered before they are introduced as solutions.

There can even be unintended and unexpected consequences of working more naturally in the vineyard. For example, compost heaps that are not managed properly can release methane, which is a far worse greenhouse gas than carbon dioxide. And travel for marketing and selling wine is a carbon footprint that is rarely considered.

Two inputs that wineries will find hard to avoid are those for packaging wine (dry goods) and transporting wine. If a winery is to be truly carbon neutral, it should cover the entire process from grape to wine glass and cannot ignore the transportation chain to the consumer.

What can the wine industry do to lessen the carbon footprint of wine? The chief culprits seem to be transportation and packaging. While it may not be possible to eliminate transportation, changing the way wine is packaged is the easiest way for the wine industry to reduce its carbon footprint. This can be done by bulk shipping and bottling in market, or lightweighting glass bottles (average bottle weight is currently around 500g/1.1 lb, whereas the lightest 75cl bottle weighs around 300g/0.7 lb), or using alternatives to glass such as bag-in-box, PET bottles, pouches, or cans.

TORRES: A LEADER FOR CLIMATE CHANGE ACTION IN THE WINE INDUSTRY

Another area of research has been carbon dioxide capture during fermentation, an area that has been pioneered by Familia Torres.

Miguel Torres has been taking climate change very seriously. Since 2007, under his guidance Torres has invested 15 million Euros (nearly 18 million USD) in renewable energy.

In April 2019, he hosted a Climate Change Course in Vilafranca de Penèdes, where a group of speakers addressed the topics of climate change and carbon neutrality. Torres explained that out of the Spanish Federation of 700 wineries, only 20 have joined the Wineries for Climate Protection organization that was established in 2011 with the idea of establishing a protocol for reducing carbon dioxide emissions. A small investment for wineries. "Still people don't see the urgency of the change we need," he says. But he adds that for the first time Spanish politicians have begun addressing climate change, and finally the government has allowed people adopting solar power to sell electricity back to the grid.

Between 2008 and 2018 Torres reduced the carbon dioxide emissions per bottle of wine by 27 percent. It has also purchased 6,000 hectares of land in Patagonia, South America, which Torres says cost the same as a luxury apartment in Barcelona. And on this property they have started planted trees to reduce their carbon footprint further. "It is a little grain of sand to help with global warming," says Torres.

Climate change is a big concern for him. He says that their viticulture has had to change in order to delay maturation. "In the 1980s, the idea was to advance maturation," says Torres. "Now it is the contrary: we want to delay maturation, or else we will have to harvest in August. We have the sugar then but not the phenolics: they are still maturing."

Planting density is also changing in the face of a warming climate. "In the 1990s, we wanted to get closer. Now we are going back again to wider spacing," he says. And they don't deleaf the fruit zone anymore. "We prefer to keep this microclimate and delay maturation." Hail is a big problem in climate change, especially in

Above Exytron's system for capturing carbon dioxide. This system shows promise for sequestering the carbon dioxide produced during fermentation, and is pictured here at the Torres winery in Spain, where they are looking at ways of reducing their carbon footprint.

Above A flower-petal-shaped solar array at the new winery of Thibault Ligier-Belair in Bourgogne. This array is able to track the sun's movement during the day.

mountain areas. Nets protect but they can also delay maturation by about a week. And gobelet training (bush vines) is also coming back, because this is well adapted to warmer climates.

Torres began buying land at high altitude ten years ago, at 1,000m (3,281ft). "The temperature is around 10°C [18°F] less per day," he explains. "We also have land at 1,200m (3,937ft). It's still a bit early." In an attempt to reduce carbon footprint, they have invested in photovoltaics (a system based on solar power) and biomass, and they have been substituting diesel cars with hybrid and electric cars. They have also tried to reduce the weight of bottles and work with alternative packaging with a lower footprint.

Carbon capture is a focus for Torres. "The vines are taking carbon dioxide from the atmosphere," he says. "Ninety-nine percent of the time farmers burn the prunings. Then when we ferment, carbon dioxide goes up again. The idea was that we could capture carbon dioxide from the fermentation." The experiments started in 2016, and a number of methods were tried. Finally, Torres collaborated with a company called Exytron, whose method involves transforming carbon dioxide to methane, which can be used as a biogas.

Exytron is a young company founded in Germany in 2013. Their goal is to convert hydrogen to methane using carbon dioxide from combustion. The methane is burned, and the carbon dioxide is then recycled by the process. In the case of wine ferments, carbon dioxide can be captured directly. The result is a fuel source with no carbon dioxide emissions, nor any nitrogen oxides. It relies on the Sabatier reaction first detailed in 1902, in which carbon dioxide plus hydrogen becomes methane and water. "The future is to decarbonize," says Torres. They hope to be able to capture ten percent as a first step, but the cost is high. "This costs five million Euros [about 6 million USD] plus more photovoltaic panels to power it," he says. "Hopefully we will get some subsidies from Brussels, or else the board is going to kill me. Unless we have more taxes on fossil fuels it will be difficult to do more."

CLIMATE AND WINE

As we have discussed, grapevines are extremely fussy about where they are grown. For the Eurasian grapevine *Vitis vinifera* to crop successfully and produce grapes capable of making high-quality wine, it needs to be planted in locations with fairly tight climatic boundaries. Typically, in the northern hemisphere these conditions exist between 28 and 50 degrees latitude, although factors such as proximity to water bodies, altitude, and continentality can create exceptions. More specifically, each grape variety has different requirements, and in order to excel must be grown close to the cooler limits of where it can be grown. The best wines are made where the vines only just attain full ripeness toward the end of the growing season. Great wines are made in the margins.

CLIMATE INDICES

How can climate be captured in numbers? It turns out to be quite tricky: climate is so difficult to sum up in a single metric. The most widely used measure of a wine region's (or potential wine region's) climate is using growing degree days (GDDs). This measure sums daily average growing season temperatures above 10°C (50°F), an approximate level that is the baseline for grapevine physiology to be active.

The most famous index for assessing wine-region climate is the Winkler scale, which separates viticultural areas into five zones according to these GDD heat-summation statistics. It is based on work by famed California viticulturists Amerine and Winkler in the 1940s, and has proved quite useful, despite its simple premise. The first bracket of the five on the scale has been divided into 1a and 1b, as the wine world went more cool-climate.

There are some limitations of a strictly thermal index like this. One of the problems is that each region usually shows quite a range of microclimates, and the official GDDs depend heavily on where the weather stations are located. To have a single figure for a whole region is not sensible. As an example, in the Gualtallary appellation in Argentina's Mendoza region there is a span of Winkler Zones 1–3 just within this one geographic indication. Also, the cut-off base temperature of 10°C (50°F) is not

strictly accurate. For vines, the base temperature depends on the variety and varies from around 4–7°C (39–45°F). The Winkler scale has no upper cut off, and this is a problem: above a certain temperature there is a point when the vine shuts off and phenological development stops. Other problems include the effect of day length (which varies with latitude) and also the differential between day and night temperatures. Soil water levels will affect the way that the vine develops: in more humid regions, the ripening capacity is lower than the thermal index would suggest.

This has led some to revise the concept of GDDs for them to make more sense. In the 1970s, Huglin developed his heliothermal index (HI). This takes day length into account, but otherwise is essentially a thermal index, like Winkler's. A more complex index has been devised by Tonietto and Carbonneau, which they called the Geoviticulture Multicriteria Climatic Classification (MCC). This assessed climate along three dimensions, with a cool night index (CI), a dryness index (DI) and an HI. This is more complex but captures more useful information about the climate of the region. Australian John Gladstones has worked on a lot of wine region climate indices, and he recently introduced the concept of Biologically Effective Degree Days (BEDD). Here, he refined the Winkler GDD concept by capping temperature at the level where the extra degrees are not useful to the vine (19°C/66°F), and adding in some other biologically relevant modifiers. Some selected average GDDs for wine regions worldwide, from various sources.

When it comes to cool climate, is the concept of GDDs useful, even in a refined form? The main use of climatic indices is in thinking of which potential sites could be viable for viticulture, and then for deciding which varieties to plant.

VITICULTURE AT THE EDGE

Before we even get to the issue of whether a region is warm enough to ripen specific grape varieties properly, there are certain climatic

Location	GDDs °C	GDDs °F
Kent, UK	850	1562
Nova Scotia, Canada	860	1579
Launceston, Tasmania	932	1709
Central Otago, NZ (Queenstown)	945	1733
Reims, France (Champagne)	982	1800
Geisenheim, Germany	1095	2003
Dijon, France (Burgundy)	1140	2084
Niagara Peninsula, Canada	1178	2152
Colmar, France (Alsace)	1214	2218
McMinnville (Oregon)	1245	2273
Bordeaux, France	1351	2464
Coonawarra, Australia	1400	2553
Bordeaux, France (Merignac)	1627	2961
St. Helena (Napa, California)	1816	3302
Stellenbosch, South Africa	1945	3533
Fresno, (California)	2584	4684

constraints that can limit viticulture. The first is early season frost. Some sites are particularly frost prone, and this can determine where viticulture is feasible. Some frost risk can be ameliorated by installing frost fans in the right places, or by the use of overhead sprinklers, or by chartering helicopters. However, helicopters and frost fans only work where there is an inversion layer, with slightly warmer air lying above the freezing air at ground level. In the absence of this inversion layer, they are of no use. For overhead sprinklers to work they must coat the new growth with water, which can then freeze and protect the shoots by the latent heat released on freezing. This requires plenty of water and adequate pumps.

Then there are winter lows. Once temperatures drop below around -18°C (-0.4°F), *Vitis vinifera* is at risk of winter damage. This is a very real problem in Canada and parts of Washington State. Vines are less at risk if they have time to adapt for lower temperatures at the end of the growing season before the extreme lows hit. And some varieties are much more resistant than others, as winegrowers in Canada's Niagara Peninsula keep finding out every decade or so when there is a very cold winter. In Prince Edward County, in southern Ontario, and in the wine regions in Québec, the winters are so cold that the vines have to be trellised close to the ground, and at the end of the season next year's canes are tied down to the fruiting wire and the vine is hilled over. The real pain is unearthing the vines at the beginning of the next season: because they have started to grow again, this must be done with the same care of as an archeologist working on an ancient site. The use of geotextiles as a cover is becoming increasingly popular in these regions as it avoids the need for hilling over and allows for a greater survival of primary buds.

For cool-climate wine regions to establish themselves, there is a need for pioneer winegrowers prepared to take some risks. New Zealand's Central Otago is a relatively new region, which has its roots in the works of a small group of experimenting pioneers back in the early 1980s. It was not possible to get finance for new vineyards because government advisers had decreed the region unsuitable for viticulture as it was considered too marginal to ripen wine grapes consistently. The first commercial wine was released in 1987, and now this is a highly successful region, with Pinot Noir its calling card. Yet the GDDs for Central Otago, at under 1,000°C (1,832°F), suggest that you'd never be able to ripen Pinot successfully here. Central Pinots are certainly ripe, though, and are some of the country's most sought after.

Two of the coldest wine regions on the globe are beginning to make a name for themselves with sparkling wine: the Atlantic province of Nova Scotia, in Canada, and southern England, in the UK. Both of these regions owe their newfound success to a pioneering producer in each case. In the UK, it was the commitment by sparkling-wine specialist Nyetimber to Chardonnay and Pinot Noir, as opposed to the earlier-ripening varieties and hybrid crosses that other producers were backing. This led to the emergence of serious traditional-method sparklings that convinced other producers that it was a viable proposition. Nyetimber's first wines, which appeared in the early 1990s, created quite a storm. Now, the English sparkling-wine industry is continuing to grow significantly and is attracting international attention. The chief challenge is the cool, uncertain weather during flowering that can impact on yields, and the occasional cool vintage, such as 2012 where the grapes never get properly ripe. But the hope is that with the modest warming trend that English vineyards have experienced over the last two decades, these too-cold vintages will become rarer.

Nova Scotia's sparkling champion is Benjamin Bridge, who based in the Gaspereau Valley. In a region where hybrids such as L'Acadie Blanc are widely grown (admittedly with some success), Bridge is making great progress with traditional Champagne varieties. A combination of careful viticulture and privileged sites, together with savvy winemaking, is resulting in traditional-method sparklings of world-class quality. Both the southern England and Nova Scotia are properly marginal. The extremely long growing season and cool temperatures result in grapes with perfect analysis for sparkling wine (although sometimes the acidity is a little too high), with proper physiological maturity.

Even established wine countries are busy exploring cooler climates. In Chile, when Casablanca was first developed in the 1980s, this was seen as a brave move into a cool climate. But Casablanca now looks positively warm compared with new vineyard developments in Leyda, Elqui, and coastal Aconcagua and Colchagua.

Johann Meyer, working in the Swartland in South Africa, uses cold units as a measure of where it is best to plant vineyards: he is looking for the coolest climate he can find in the warm climate of the Western Cape. "When I look at planting vineyards I look at cold units during the year," he explains. "The main area they use this is in fruit planting. If you want to plant an apple tree, you need at least a minimum of 800 cold units. If there are apples, then I can plant Pinot Noir and Chardonnay. They need the same kind of cold units." He gives the example of Hamilton Russell's vineyards in Hemel-en-Aarde, which get 780-820 cold units per year. "I get 1,100 per year," he says, talking of his new vineyard plantings on the fringes of Swartland. "My average daytime temperature here is 2°C [3.6°F] colder than Hamilton Russell. Altitude is responsible: for every 100m [328ft] in altitude you drop 1°C [1.8°F]. The highest vineyard here is 820m [2,690ft]."

5 Pruning, trellis systems, and canopy management

The grapevine is a something of a freeloader. It has adopted a growth strategy where it can't be bothered to support itself and relies on others instead. In nature, plants are in competition for two sets of resources, those from the earth and those from the sky. In most environments the latter struggle is the key one, the struggle for enough light to drive photosynthesis and hence food production. Trees often win this battle by getting their leaves 9 m (30ft) or more from the ground, but to do this they have to spend years slowly building a woody trunk with enough mechanical strength to support this elevated growth habit. Grapevines, like other climbers, have seized the opportunity to make the most of this third-party effort. They realized that by climbing they could save themselves the trouble of developing a self-supporting stem. Not having to develop supporting girth permits rapid growth, so grapevines are experts of growing up other plants until they break through to uncompeted-for light on the outside of the canopy of their hosts. They are structural parasites.

The growth habit of the vine is finely tuned for this lifestyle. Shoot structure is simple, with each node having the capacity to produce tendrils or flower buds opposite each leaf. Gripping to the host plant via these tendrils, the shoots grow rapidly toward the light, seeking the gaps in the canopy. Where the vine breaks through to sunlight, tendrils are discarded in favor of flowers, resulting in fruit production. At the other end of the vine, the roots are capable of growing deeply, eking out water and mineral resources in competition with the preexisting root system of the host plant.

The science of viticulture attempts to manipulate vines to get them to produce good yields of high-quality fruit. It takes into account the natural growth habit of the vine and adjusts it to suit the context of the vineyard. Because vines are climbers, this usually involves some means of support for the freeloading grapevine to grown on. It is an area where good controlled scientific experiments are rare, in part because of the immense difficulty in doing them. The way vineyards look today is a reflection of tradition, trial and error, guesswork, specific environmental constraints, and convenience. As a result, vineyards across the world's wine growing regions differ markedly in their appearance. This chapter sets out to discuss some of the scientific considerations that shape the way that vines are pruned, trellised, and managed.

VITICULTURAL GOALS
The holy grail of viticulture is to get high yields of high-quality grapes with minimum cost in terms of vineyard labor and inputs. Almost invariably, a compromise is involved: sometimes yield must be sacrificed for quality; sometimes it's the other way around. Good viticulture also takes into account the economic objectives of the wine that's going to be made from the crop. Vineyards are therefore managed to produce grapes of appropriate quality and at the right yield, for the right cost.

If you are starting a vineyard from scratch there are several key choices that need to be made. These include choosing the appropriate variety, vine spacing, trellis method, and making decisions regarding irrigation, pruning, and canopy management. Making the right choice is important, because the vines will have a productive lifespan of 20 years or more. This timescale makes innovation based on experimentation tricky, and so in areas where vineyards are already established people frequently copy the vineyard style of their neighbors. Often viticulturalists have to work within the confines of vineyards that have already been planted, in which case there is limited room for modification. In the following sections I'll give a brief overview of some of the scientific principles behind trellis systems, pruning, and canopy-management decisions, and how these all affect wine quality.

Current viticultural thinking is that the control of vine vigor and fruit-zone light exposure is the key to successful vineyard management. Vigor is an important element of viticulture. If the vine is growing actively all through the season, developing a huge, dense canopy, then the actively growing shoots will represent a powerful sink for the vine's resources that will inhibit the sugar accumulation, which normally occurs during the fruit-ripening phase. The immediate result is delayed fruit development and lowered quality, but also of importance is the effect of the profuse canopy in shading the inside-growing shoots. This is because light is of crucial importance for bud fertility. In the fertile regions of the shoots (next year's canes) grapevines have uncommitted bud primordia that can form either tendrils or flowers, and because the development of these buds takes two seasons, it is the light that the fruiting canes received in the preceding season that determines their fertility in the current year. This is understandable, and makes sense for a grapevine in the wild, where it will be growing on trees because it only wants to make fruit where the shoots poke out through the host canopy into the light. In addition, shaded grapes may maintain high levels of vegetal-tasting methoxypryazines, which are undesirable in most wines and are dissipated through light exposure. In support of these ideas, precision viticulture studies have emphasized that in many wine regions, the parts of vineyards that produce the highest-quality grapes are those with the lowest vigor. Excessively shaded canopies have another drawback in that they increase the risk of disease. This is because of the limited air circulation and the extended drying-out time after the canopy is wetted.

Another key to successful viticulture is getting grape flavor development (known as phenolic or physiological maturity) to coincide with sugar maturity. In warmer regions, the risk is that by the time the grapes have reached flavor maturity, the sugar levels are very high, resulting in overly alcoholic wines. This is currently a major problem in many New World wine regions. In cooler regions, flavor maturity often occurs at much lower sugar levels and the challenge can be getting the grapes ripe enough in terms of sugar before the grape loses its photosynthetic capability or the fall rains set in.

With trellis systems and canopy management there is no one-size-fits-all solution. Viticultural methods have to be adapted to local conditions, and while the general principles remain the same, factors such as soil fertility, climate, water availability, grape variety, and the skill and availability of vineyard workers must be taken into account in the consideration of the most suitable management choices.

PRUNING

Pruning is an intervention that aims to improve vine fertility, encourage optimum canopy development, and regulate crop load in line with the quality objectives of the grower. Vine pruning can seem a little complex for those unfamiliar with it, so here is my attempt to present a digestible introduction.

There are two different styles of pruning, both widely used, known as cane pruning and spur pruning. Cane pruning involves selecting one or two (rarely more) shoots from the previous season's growth and cutting them back to between six

Above Cane pruning at Hambledon Vineyard in Hampshire, England. The vine is pruned back to two canes and two replacement spurs. Here the canes have been bent down a little. This promotes even bud burst, as the buds at the end of canes tend to take off fastest.

and fifteen buds. These then form the basis for the following year's growth when they are tied down horizontally. A renewal spur is also left for generating new canes. Typically, with cane-pruned vines the only permanent vine growth is a vertical trunk. The actual practice of cane pruning is more challenging than spur pruning and is usually employed with varieties that have low fruitfulness in basal buds, and where the vineyard workers are up to the task. Results can be very good. It's ideal for cooler climates and certain varieties where basal buds have low fruitfulness. Single Guyot is an example of cane pruning where one cane is laid down on the fruiting wire, with a short replacement spur also left. Double Guyot is where there are two canes, laid down on each side of the vine, again with a short replacement spur. In New Zealand's Marlborough region, cane pruning is used because Sauvignon Blanc has low basal bud fertility. Some vineyards looking for higher yields will have four canes laid down, two on each side, on different fruiting wires.

Spur pruning involves cutting the previous season's growth back fairly drastically to just a few (up to five, but more normally two or three) buds. These will be borne on a more substantial permanent vine structure, usually consisting of a trunk plus horizontal cordons. Spur pruning is technically much simpler and requires less skill on behalf of vineyard workers. It can be partly mechanized by using mechanical pre-pruning (sometimes called "bushing") that may then be tidied up by hand.

Minimal or mechanical pruning is a relatively recent development. This involves no real pruning, just cutting the vine's growth back rather crudely using mechanical means. It makes a mess of the vineyard, but proponents claim that after a couple of years the vine gets into balance and produces good yields, with small bunches of grapes all over the canopy, rather than clustered in a fruiting zone. This is therefore only compatible with mechanical harvesting. It is used in situations where manual vineyard labor isn't practical or economically feasible, and only really works in warm climates. A variation on the theme is mechanical cutting back of vine growth, which is then tidied up by vineyard workers manually. This is evidently only going to work for spur-pruned vines.

Pruning has recently come under the spotlight because of its proposed role in grapevine trunk diseases. This is discussed in Chapter 6, see page 89.

TRELLISES AND CANOPY MANAGEMENT

There exists a confusing array of different styles of trellis, some traditional artifacts, no doubt, but others each with their own advantages in specific situations. Trellis systems are a vital part of the canopy-management tool set.

Canopy-management techniques are aimed at achieving optimum leaf and fruit exposure to sun, while reducing the risk of disease and pushing the quality to yield ratio as far as possible. Open canopies help prevent disease in two ways. They allow better spray penetration and also better air circulation, with faster drying-out times. Such strategies aim to get the vine into some sort of balance, and they have been particularly successful in situations of high vine vigor, often caused by fertile soils and irrigation. The modern canopy-management techniques that involve, for example, split-canopy trellis systems, such as the Smart-Dyson, are not, however, of much value in low-vigor sites, such as the major vineyard areas in the classic Old World regions.

Richard Smart's work has been particularly influential in this area. He dubs the traditional canopy-adjustment techniques of trimming, shoot thinning, and leaf removal in the fruit zone as "band-aid viticulture," because they are interventions that have to be reapplied annually. His solution is to alter the trellis technique to increase canopy surface area and decrease canopy density—a once-only intervention. In particular, the use of high vertical-shoot positioning (VSP) systems and divided canopies (such as the Smart-Dyson and Scott-Henry trellis systems) have been effective means of getting highly vigorous vines into balance. The basis of this work is to manage the vine vigor, achieving optimum leaf-to-fruit ratio. Smart considers this to be as important a consideration as the traditional vineyard currency of yield.

I asked viticultural consultant David Booth, who worked in Portugal (and, sadly, died prematurely in 2012 at the age of 47), about his views on canopy management. "I think it is probably one of the most important tools we have," he responded. "But I do have a very broad definition, much more than just putting up technically advanced trellis systems. My definition encompasses a range of vineyard-management practices, including winter pruning, shoot thinning, shoot positioning, leaf thinning, and hedging. The skilled viticulturalist should be able to look at the soil profile (before planting) at any particular site and anticipate future vine vigor. Then they can make a series of decisions about trellis system, spacing, and rootstock. A high VSP is probably the easiest to manage and

I reckon this should be the natural first choice. The more advanced trellis systems are for when the anticipated vine vigor is so high that you have doubts about your ability to accommodate the growth within the VSP, or several years after planting you realize you have blown it and underestimated vine vigor and need to modify the existing trellis. My first choice for divided-canopy systems is Smart-Dyson, since it is easy to manage, does not require wide rows and is easy to machine harvest." Booth adds, "A key factor to think about in canopy-light environment is not just sunlight striking bunches, but also leaves shading other bunches, which has negative implications for wine quality, for reasons that are not understood but probably have a lot to do with potassium balance."

There are two dogmas in viticulture that are worth addressing here. One is that reduced yield equals higher grape quality (and vice versa); the other that old vines produce better wines. The scientific bases behind both of these assertions are not entirely clear. Evidently, it is possible to lower grape quality by means of excessive yields. And in classic Old World regions, which are typically low-vigor sites, reducing yields by pruning canes short does have the effect of raising quality, to a degree. But in higher vigor, irrigated vineyards in warm regions, pruning short will not result in a vine that is in balance, and no quality gain is likely to be seen. In such high-vigor situations, moving to a split-canopy system often has the effect of bringing the vine into balance, raising yields, and improving quality at the same time. The second dogma is that old vines produce better vines. It is repeated so often, the suspicion is that there must be some truth in it. If it is indeed the case, what is the scientific explanation? One suggestion is that it has to do with the amount of overwintering perennial wood on the vine. Some researchers have noted that training systems with more perennial wood, and thus more carbohydrate storage area during the dormant period, produce better wines. Older vines tend to have more perennial wood and this could be to their advantage. However, a more likely explanation is the one offered to me by David Booth. "Young vines are harder to manage because they are less buffered against any environmental stresses. But as you know, they can give great quality, probably because they are naturally low vigor (small root system) and have good leaf and bunch exposure. Old vines are also naturally low in vigor, due to wood disease and exhaustion of nutrients. I reckon the problem is more in the middle years, especially in high vigor soils with inadequate trellising systems—then you get the classic shading problems." I asked him whether it is possible to increase yield and maintain or even improve quality by viticultural interventions, such as canopy management. "Sure, but only really in the case that I have just mentioned of the middle-aged vigorous vine on the high-capacity site," replied Booth. "These sites are more common than you might think. Look for the small yellowing leaves in the inside of the canopy, as an indication of leaf shading. A full-on trellis conversion may be the solution, but I tend to work first with competitive cover crops, nutrition, irrigation management, leaf pulling, and shoot thinning. As you might have figured out by now, there is no silver bullet, just a raft of tools that often need to be used in conjunction."

Some common viticulture terms

Basal leaf removal Basal leaves are removed to expose the fruiting zone, allowing access to light and encouraging air circulation, which prevents disease.

Cane The stem of a grapevine that is one season old and has become woody, and which can either be cut back to spurs (up to four buds) or canes (typically six to fifteen buds) for the following season's growth. Canes are also sometimes called "rods."

Cordon The woody framework of the vine extending from the top of the trunk. A cordon-trained vine has a trunk terminating in one or more cordons, which are then spur pruned.

Head training Where the head of the trunk is pruned to either spurs (gobelet system) or canes (such as the Guyot system, which was named after its inventor).

Hedging Also known as shoot tipping, this involves cutting back excessive growth at the top and sides of the canopy midway through the growing season. The aim is to leave enough leaves to ripen the fruit, while preventing excess growth that will lead to shading and competition for resources with the fruit. A balanced vine will typically have two fruit clusters and fifteen leaf nodes on each shoot.

Shoot Green growth arising from a bud.

Spur A short cane that has been cut back to between one and four buds to provide the following season's shoots.

Trunk The main, permanent vertical growth of the vine, which supports the canes or cordons. Grows in girth only.

Above Freestanding vines at the Lopez vineyard in the Cucamonga Valley near Los Angeles. Owned by the Galleano family, this was planted mainly with Zinfandel in 1918. These unirrigated vines sprawl over the sandy soils.

Above Classic bush vines in the Languedoc, France. These unsupported vines are idea for hot climates, where their compact canopies don't need a lot of water and give a little shading (but not too much) for the grape clusters.

Above Chardonnay in the famous Montrachet Grand Cru in Bourgogne, Franace. These vertically positioned canopies grow from a low-fruiting wire and are hedged and trimmed to a relatively low height.

Left A version of pergola in Trentino, Italy.

Above A pergola system in Yamanashi Prefecture, in Japan. Here, the bunches of Merlot are covered with waxed-paper hats. This is to protect them from the abundant growing-season rainfall.

Above In this classic cordon, the vine has two permanent arms with canes pruned back to short spurs.

Above Geneva double curtain training system at Ancre Hill Vineyard in Wales, in the UK.

Above A lyre system, Mornington Peninsula, Australia.

Examples of different training systems

Cordon de Royat Simple and effective spur-pruned system, usually with a unilateral cordon spreading from a low trunk. Variations include a double cordon.

Dopplebogen The "double bow" system for Riesling vines common in the Mosel region of Germany, where each vine is singly staked and two canes are bent around into a bow shape. Single-staked vines are also found in France's Northern Rhône, called *sur echalas*. This sort of system is adapted for steep slopes.

Éventail French term for "fan," this is a cordon system with a number of arms arising from a short trunk, each bearing a short cane. This is popular in Chablis and also used in Champagne.

Geneva double curtain A rather complicated split-canopy system, with cordons grown high up on two parallel horizontal trellis wires, with the shoots bending down. A variation is the lyre system, which is relatively common in Austria, but here the split canopies have upward-growing shoots, and angle outward slightly. Both are hindered by the fact that they require wider rows and complicated trellis systems.

Gobelet An old, and probably the simplest, training technique, where the spurs are arranged around the head of a trunk or short arms coming from the top of the trunk. This is only really used in warm, dry climates in low-vigor situations. It doesn't need a supporting trellis but shading of fruiting zones can be a problem (though it could be argued that the dappled light the low-vigor canopy provides is ideal). The style is popular in the Mediterranean regions. In the New World, this is known as the bush vine; in Italy it is called the *alberello*.

Guyot One of the most popular cane-pruned systems, with a single or double cane layered horizontally from the head of a trunk. One or two renewal spurs are also left. Simple and effective, it is particularly suited to classic Old World, low-vigor vineyards.

Lyre A split-canopy system, this has two canopies facing each other forming a "V" shape from a single cordon. It is complicated but can give good results.

Scott Henry A split-canopy trellis system where the shoots are separated and divided into upward- and downward-growing systems, held in place by foliage wires. It is useful for high-vigor situations and is suitable for both spur and cane pruning. The advantage is lower disease pressure, improved grape quality, and higher yields. Looks like a wall of vines in practice, growing from the ground to about 2m (6 ft) in height.

Single-wire umbrella sprawl Commonly found in Australia, the shoots sprawl in an umbrella fashioned from a single wire. It looks untidy, but the fruit is shaded in dappled light, preventing sunburn.

Smart-Dyson Developed by John Dyson and Richard Smart, this is a variant of Scott Henry trellis, with curtains trained up and down from just one cordon. Popularized by the influential work of Smart.

Sylvoz A high-yielding system sometimes used for making sparkling base wines, with many downward-pointing "hanging" canes.

Tendone The Italian term for the arbor or pergola system common in parts of Italy, Portugal, Spain Argentina, and Chile. The vines are trained high off the ground on wooden, metal, or stone frames. They look attractive, and yields can be heroic, but fruit shading is a problem and quality suffers as a result. They are hard to work, too. This trellis system replicates most closely the growth of the vine in the wild.

Vertical shoot positioning (VSP) Widely adopted system where the shoots are trained vertically upward in summer, held in place by foliage wires. Leads to relatively tall canopies and is suitable for mechanization. VSP is good option for most sites.

6 When things go wrong

Growing grapevines can be challenging. Because of their susceptibility to disease and their vulnerability to insect pests, vines need spraying. And one introduced insect pest in particular—phylloxera—almost eliminated viticulture back at the end of the 19th century. Powdery and downy mildew have been a scourge of vineyards worldwide for 150 years and are the main reason vineyards need frequent chemical help, with its attendant environmental impact. And in recent years a new problem has become apparent in the vineyards of the world: the rising incidence of grapevine trunk diseases. In this chapter I explore some of the ways things go wrong in vineyards, and look at old and new scientific approaches for combating them.

POWDERY MILDEW

Anyone who's ever tried to grow grapevines without spraying them will probably have come across powdery mildew. It is the biggest disease of grapevines, is found everywhere in the wine world, and all *Vitis vinifera* varieties are susceptible. *Erysiphe necator* is the scientific name for grapevine powdery mildew. It was originally known as *Oidium tuckeri*, and some people still refer to it as oidium, especially in France. You might also see it referred to as *Uncinula necator*, but *E. necator* is the standard name.

E. necator hails from the USA. Here, native grapevines have had a very long time to coevolve with this fungus and develop a degree of resistance to it. But then, in 1845, it found its way to the hothouse of an English gardener. Edward Tucker, in Margate, in Kent, noticed it on one of his vines. He consulted with M.J. Berkeley, a Bristol-based botanist, who examined the fungus, and then named it *Oidium tuckeri*. Tucker also had a look under a microscope and decided that this disease was very similar to peach mildew. Back in 1821, an Irish gardener called John Robertson had successfully used sulfur to control peach mildew, so Tucker tried it too, and it worked. Robertson had mixed sulfur with soap, whereas Tucker mixed it with lime. Despite the successful treatment, powdery mildew spread across the UK and to France. Before long, it was a major problem across the vineyards of Europe. It was first spotted in French vineyards in 1847, and then in Italy in 1849. In the early 1850s, it caused massive problems economically, as winegrowers learned to get to grips with the best ways of applying sulfur as a remedy. Since then, it has been ever present in the vineyards of the world.

There are two ways that it spreads. The first is by structures called conidia, which are asexual propagules dispersed by the wind. The fungus overwinters as a mycelium (a mass of fungal hyphae, which are like fine tubes). Often, the mycelium will overwinter in a developing bud. When the vine starts growing, the bud opens and the entire structure that grows out will be infected by the fungus. This is called a flag shoot. After a while, all the green parts of this shoot are covered by what looks like a white powder: these are the conidia, and they infect the rest of the vine. The second way is by what are called cleistothecia. These are the sexual stage of the fungus and are particularly dangerous because when the fungus has sex it jumbles its genes around and resistance

Above Powdery mildew on a cluster. This is one of the main disease threats to winegrowers worldwide, and, unlike downy mildew and botrytis, it isn't associated with damp conditions.

to any fungicide is much more likely to occur. The cleistothecia are black spherical bodies and they contain four–six asci, each containing four–seven ascospores. They typically overwinter in the bark of the vine, then in the spring release the spores that germinate and restart the fungal disease. Cleistothecia are not present in all countries where powdery mildew is found.

All the green parts of the vine are susceptible to powdery. The conidia or the ascospores germinate to produce tip-growing filaments called hyphae. Along their length, they produce structures called appressoria, which pierce the epidermal cells. Haustoria are then formed inside the plant's epidermal cells, which farm nutrients and enable the fungus to carry on growing. In turn, the fungus then develops new conidia or cleistothecia and spreads itself further. When powdery mildew affects the grape berries, it damages the epithelial layer. Thus while the berry continues to develop, the epithelium doesn't and the berry splits.

HOW TO DEAL WITH POWDERY MILDEW

Interestingly, while most people associate dampness with fungal diseases, humidity isn't a vital factor in the spread of powdery mildew. It is what is known as a xerophytic plant pathogen:

the conidia are adapted to germinate under dry conditions, and rain can actually reduce germination. But there is a twist here. *E. necator* doesn't like UV light, so damp, shady conditions can work in its favor. There are currently trials looking at using UV lights to treat against powdery mildew. They seem to be effective, but the UV has to be applied during the dark, and the challenge will be to get a lighting rig that is robust enough to use in a vineyard.

This is one of the reasons that canopy management is a tool for mildew management. Opening up the canopy, and particularly the fruit zone, is protective, because sunlight reduces disease pressure. It also enables penetration of the leaf and fruit surfaces by fungicides, which is especially important when we're talking about sulfur, which is a contact fungicide and needs to be applied evenly across all surfaces to be effective.

The main treatment is still sulfur, which is usually applied as a wettable formulation. It's a contact fungicide that needs to cover all the green areas of the vine and must be reapplied at regular intervals. Because it has several mechanisms of action, resistance doesn't develop. There are also systemic fungicides that are more targeted to specific mechanisms of action. For this reason, these need to be alternated by fungal class, to avoid resistance developing. Most of these are preventive, but some also have eradicant action.

There is a new organic product with eradicant action, HML32. This is a potassium-based liquid soap, and when it is applied with a copper fungicide (such as Nordox) and potassium bicarbonate, it is quite effective at knocking back an existing powdery infection.

WHAT ABOUT RESISTANCE?

Twelve different genes have been identified in American and Asian native grape species that confer resistance to powdery mildew: Run1 and Run2 (where Run stands for resistance to *Uncinula necator*) and Ren1-10 (where Ren stands for resistance to *Erisyphe necator*). Can these be bred into vinifera? Attempts are well under way, and there exist a number of hybrids that have a high percentage of vinifera genes but also good resistance to powdery mildew (see chapter 7, page 96). Whether or not the wine world is ready for new varieties like this remains to be seen, but unless the world embraces transgenic (genetically modified) grapes varieties, then this could be the only way that the excessive spraying of vineyards can be avoided.

DOWNY MILDEW

After powdery mildew, downy mildew is the major disease concern for winegrowers, especially those who farm in more humid regions. But despite the similarity in the commonly used name for the disease, while powdery mildew is caused by a fungus, downy mildew isn't (even though it's common to refer to both maladies as "fungal disease"). Its causal organism is a peronosporomycete, *Plasmopora viticola*. (It is sometimes also referred to by another name, Peronospora.) Previously known as oomycetes, peronosporomycetes are a strange group of organisms related to brown algae and diatoms, and they are insensitive to many fungicides. *Phytophora infestans*, responsible for the potato blight that led to the deaths of a million Irish in the mid-19th century, is part of the same group of organisms. Unlike fungi, they do not have chitin in their cell walls. Distinctively, they produce motile zoospores each with two flagella, which are asexual propagules that actually swim. That's why they like humid conditions to spread! There is also some evidence that there may be more than one species involved in grapevine downy mildew.

If conditions are right, downy mildew spreads rapidly. It is highly destructive, causes leaves to die, and—early in the season—it can also affect the fruit. The main effect is to reduce the photosynthetic capacity of the vine, affecting ripening and also the storage of carbohydrate for the following season.

Just as with powdery mildew, downy mildew is another disease that originated in the USA. The first description of it was in 1834, when it was called *Botrytis cana*. In 1848, it was renamed as *Botrytis viticola*. It was first identified in France in 1878, probably arriving with cuttings of American vines, and three years later, it was widespread throughout France and Algeria. In 1888, it was renamed *Plasmopora viticola*. The damage caused by downy mildew is highly dependent on the weather, but it can be devastating. In 1915, for example, 70 percent of the French grape crop was lost to it. In a recent study from Trentino, in Italy, looking at the effectiveness of mildew treatments, the authors found that the untreated control plots lost 60–95 percent of their leaves, and 20–100 percent of grape bunches were affected.

Infection occurs through the stomata, which are small closeable openings on the surface of green tissues of plants that allow for gas exchange. *Plasmopora viticola* overwinters in the form of bodies called oospores. In spring, they germinate to release macrosporangia. These are spread onto vine leaves by rain splashes. In turn, they release motile zoospores that lack cell walls and have two flagella. They head to the stomata, causing the primary infection. There, they lose their flagella and undergo a process called encystment, developing a cell wall. They form hyphae that go through the stomatal hole. If forms feeding bodies called haustoria that take nutrients from the vine's cells. Then, when it is humid, they produce sporangiophores and sporangia, which cause secondary infections, spreading the sporangia via water splash or wind. These, in turn, liberate new zoospores.

The first signs of infection are what are known as oil spots on the upper surfaces of leaves. Turn the leaf over and you will see white sporangiophores and sporangia appear as a downy mass on the underside. These oil spots go brown, dry up, and the tissue dies. The risk of downy mildew can be predicted and the first

Above Downy mildew. On the top of the leaf (left) it appears as oil spots, with a white growth on the underside (center). The third picture (right) shows the aftermath of a mildew attack.

spray applied according to what is called the 3–10 spray strategy: when the average temperature is more than 10°C (50°F), when there is more than 10mm (0.4in) rain in 24 hours, and when the shoot length is more than 10cm (4in). Alternatively, more sophisticated forecast models can be used.

BORDEAUX MIXTURE AND OTHER FIXES

The chemical fix, still used today, was discovered by Pierre-Marie-Alexis Millardet (known simply as Alexis Millardet), who came up with the solution to this crisis, which he discovered fortuitously. He was an important figure in French viticulture as it navigated the triple crisis of the two mildews and phylloxera. In October 1882, Millardet was in the vineyard of Château Beaucaillou in the St-Julien region of Bordeaux and he noticed something strange. Rather than leaves devastated by downy mildew, he saw vines with a healthy canopy. The leaves had been covered with verdigris—copper sulfate mixed with lime—in order to deter thieves. It gave them a bluish green color, but it also stopped the mildew in its tracks. Over the next two years, Millardet revised this treatment, which became known as Bordeaux mixture. The recipe is quite simple. Mix 100l (22 gallons) of water and 8kg (18lb) of copper sulfate. Then make a milk of lime, by dissolving 15kg (33lb) of rock lime

with 30l (6.5 gallons) of water. Just before spraying, combine the two and use it all up within a day or two; 50l (11 gallons) of this concoction treated 1,000 plants. So, to treat one hectare would take around 500l (110 gallons). This solution was effective, and variations on the theme of this Bordeaux mixture are still in use today. It is permitted under organic rules. But the level of copper used is limited, because it can cause issues with soil toxicity.

The active ingredient in Bordeaux mixture is the cupric ion, Cu^{2+}. This protects the vine against fungi, peronosporomycetes, and bacteria by affecting their membranes and interfering with their enzymic reactions. Because it affects multiple sites in the pathogen, it is extremely unlikely that resistance will evolve. It has to be present before infection occurs: it is preventive but not curative. The copper is not absorbed by the plants and eventually gets washed off into the soil. Because it is not degraded (it kills microbes) it can remain as a contaminant for a long time. Continued copper use leads to its accumulation in the soil if it isn't leached away, and it often becomes insoluble, binding to various soil components.

There is currently some discussion at an official level about eliminating copper in the EU. As it stands, the rules are no more than 4kg (9lb) per hectare per year on a rolling five-year average, having been lowered from 6kg (13lb) per hectare in 2008. In the scientific literature there are reports of historical levels of copper use as high as 80kg (176lb) per hectare per year. The current permitted levels for vineyards seem not to present any health implications for wine consumers. A study in 2013 in Bari, Italy looked at an organic vineyard where 7.5kg (16.5lb) per hectare per year had been applied in one season, and the berries—wines from it were all below the maximum residue limits for copper.

One way to get the most out of the limited amount of copper that can be applied is to spray more carefully, so that none is wasted, and also spraying using a recovery system. Modern formulations are also more effective than the original Bordeaux mixture. Micronization gives smaller particle sizes, improving coverage. It can also be combined with zeolites, a clay-like bentonite that helps stick the copper to leaves, stopping it from being washed off. One study looking at some of these formulations found that it was the amount of copper that provided the protection, and the lower rates in some formulations might not provide enough protection. They point out, though, that higher levels than are required for protection provide no additional benefit.

Potassium phosphonate is also a useful treatment for downy mildew and was for a long time allowed in organic viticulture. This is the reason for the lower limits for copper in Germany and Austria. But a while back, phosphonates were declared illegal in organic winegrowing in the EU. In a paper back in 1999, Speiser and colleagues looked at 13 farms evaluating the use of phosphonate, and then tested 53 wines from these farms. They concluded that phosphonate was an effective treatment, especially when used in conjunction with copper, but there were some residues of phosphonates in the wine when it had been used. Because of this, their opinion was that these chemicals were not compatible with organic

viticulture. There is a lot of work underway to try to prevent downy mildew without using copper for organic viticulture. One potential solution is to use compounds called elicitors, which help to prime or ramp up the vine's own defenses. This is discussed in chapter 7, see page 95.

For those not farming organically, there are a number of other fungicides that can be used against downy mildew. They include dimethomorph, phenylamide oxadixyl, aluminium ethylphosphate (fosetyl-A1), acylalanine metalaxyl, cymoxanil, and the strobilurin fungicides.

WHAT ABOUT RESISTANCE?

Twenty-seven different downy mildew resistance genes have been identified in native vines from America, China, and central Asia: these are Rpv (resistance to *Plasmopora viticola*) 1-27. Where *Vitis vinifera* (which lacks these) does show some resistance, this is mainly a post-infection phenomenon. Some vinifera varieties are more resistant to downy mildew than others: the more resistant varieties have a strong reprogramming of their defense responses that the susceptible varieties lack. As with powdery mildew, breeding programs have been ongoing, resulting in a number of new varieties with the viticultural characteristics of vinifera but also some of the resistance genes of native American and Asian vines.

BOTRYTIS

Botrytis cinerea is unusual among the fungi that affect grapevines in that it is normally a problem, but in some situations it can be positive. As "noble rot," it is responsible for beautiful, highly sought-after sweet wines. This is covered in detail in chapter 18, see page 194.

It is as a pathogen that it is best known, and it is not restricted to grapes. It is one of the major plant fungal pathogens worldwide, causing losses to fruit and crops both pre- and post-harvest. It can spread rapidly, because it is capable of producing lots of spores. If you've ever seen nobly rotted grapes being tipped into the press, you'll know this first-hand: it results in huge cloud of these spores. These can easily be spread by wind.

In vineyards, botrytis is an issue as harvest approaches, and particularly in wet vintages. As well as damaging grapes, it opens the way for other opportunist bacteria and moulds to grown in the damaged clusters. Botrytis produces a polyphenol oxidase enzyme called laccase, which is able rapidly to oxidize the various components of the grape must. This enzymatic oxidation is faster and more problematic than chemical oxidation.

There are two ways that botrytis tends to infect grape berries. The first is known as a latent infection. When the flower cap falls off during flowering, it leaves a ring of necrotic tissue at the tip of the torus, which is what the developing berry later sits on. Botrytis inhabits this dead or dying tissue, but the infection remains latent and the healthy green tissue is protected by compounds called stilbenes. But, as the berry matures, the level of stilbenes decreases, and the botrytis gets stuck in.

The second method is through dead flower caps and other debris that get colonized by the fungus. At bunch closure these can be trapped in the cluster and then later cause infections. If berries have any fissures or cracks in their cuticle, botrytis can get in.

HOW TO DEAL WITH BOTRYTIS

The first way to respond to this threat is to reduce the disease pressure by cultural means. It is best to open up the fruit zone by using leaf removal where possible, to allow any moisture to dry out on the bunch.

Removing debris from flower hats before bunch closure can also work. In the Marlborough region of New Zealand they found that one of the side-benefits of using blasts of air from a device called a Collard to shatter leaves in order to open up the fruit zone was that these air blasts also got rid of a lot of this debris. An alternative management technique is to send harvesters down the row before bunch closure to shake the vines. This achieves the same end and has proven very effective as a means of botrytis control. However, it is not widely adopted because growers don't like the idea of putting a harvester in the vineyard when the valuable crop is still on the vine a long way from being ready to pick.

Biofungicides are also a potential solution. A research group led by Pertot and colleagues looked at the effectiveness of biofungicides over a four-year period in trials in north and central Italy. They found that *Trichoderma atroviride*, *Aureobasidium pullulans*, and *Bacillus subtilis* applied at bunch closure, *veraison*, and pre-harvest, respectively, controlled botrytis quite effectively on the bunches. These three microbes seem to be working through different mechanisms, but they don't interfere with each other. A study by Calvo-Garrido and colleagues looked at a range of biological control agents in a three-season, ten-plot experiment in Southwest France. They compared four experimental bacterial strains with nine commercial bacterial products.

Of the experimental strains, *Bacillus ginsengihumi S38* was the only one that worked, giving a reduction in botrytis frequency of 35–60 percent. Several of the commercial strains worked, but not in every trial, giving 21–58 percent reduction. The most effective were *Candida sake* (45 percent), *Bacillus subtilis* (54 percent), and *Bacillus amyloliquefaciens* (58 percent). The issue with these products was the number of times they had to be applied in order to work.

There are quite a number of synthetic fungicides that are being used as anti-botrytis treatments, including active ingredients pyrimethanil, cyprodinil, and fludioxonil. Applied at 80-percent cap fall, they can protect the cap scars and then again at pre-bunch closure, they have a final chance to deal with any debris that might get trapped.

WHAT ABOUT RESISTANCE?

Botrytis is a necrotrophic pathogen, which means it feeds on dead or dying tissue, rather than actively infecting living tissue that might fight back. If it can get into grapes through small wounds or fissures in the cuticle, then it will infect them. This means there aren't specific resistance genes or mechanisms of resistance: any resistance is multifactorial, with small contributions from several physical characteristics. This makes it hard to breed for resistance. One of the major risk factors is bunch architecture. Compact clusters, where the berries are right up against each other, are bad. The inner surface of the cluster often has high humidity, and this can result in small fissures in the cuticles of the berries that allow botrytis in. Also, where the berries are in contact, the cuticles are generally thinner and have less wax. If there has been any powdery or downy mildew damage during the season—even if it is at a very low level— this can also open the door to botrytis. There are some native American vines that show resistance. These tend to have few pores on their berries, with thick cuticles, lots of wax, and thicker skins.

PHYLLOXERA

The global wine industry rests on a single, rather fragile pillar. This pillar was put in place rather hastily just more than 120 years ago, when wine, as we know it, was almost extinguished forever by a tiny aphid, phylloxera. This tiny insect, with its complicated life cycle, caused a vine plague of epic proportions, which in the space of a few decades brought the world wine industry to its knees. Salvation came from an unlikely source—the same as the origin of the problem—native American grapevines. The pillar in question is the resistance of American vine rootstocks to phylloxera and, fortunately, it has proven amazingly durable. The consequence of the phylloxera pandemic is that now almost all *Vitis vinifera* vines are not planted on their own roots. Instead, they are grafted onto resistant American rootstock.

In the Victorian craze for importing novel, exotic plant varieties, aided by the steamship, American vines began to be imported into France in the early decades of the 19th century, and by 1830 a couple of dozen varieties were growing in French nurseries. Nurserymen and vignerons keen to experiment disseminated these imported vines throughout Europe's wine regions. Over a couple of decades a catastrophe unfolded, caused by the root-feeding aphid, *Phylloxera vastatrix* (known today by scientists as *Daktylosphaira vitifoliae*). Monsieur Borty, a wine merchant in the town of Roquemaure, in the Gard, had the misfortune of going down in history as inadvertently precipitating this natural disaster. In 1862, he received a case of vines from New York, which he planted in his small vineyard. Two years later, vines in the surrounding area mysteriously began to wither and die. The malady spread. By 1868, the whole of the Southern Rhône was infected and phylloxera had become endemic throughout the Languedoc. Within the space of a decade, it had moved through France and gone global, reaching Portugal's Douro by 1872, Spain and Germany two years later, Australia in 1875, and Italy by 1879. The last French wine region to be affected was Champagne, in 1890.

At first, the outward signs (the leaves of affected vines turning yellow and falling prematurely, followed by death) gave no clue as to the specific nature of the disease. When affected vines were extracted from the ground, their roots were found to be rotten and crumbling, but with no obvious pest present. It was down to Prof. Jules-Émile Planchon, a botanist appointed by the government commission established to investigate the outbreak, to identify the culprit. He dug up roots of a healthy vine and observed clumps of minute wingless insects happily gorging themselves. Clever parasites don't kill their hosts, and in their homeland of the USA phylloxera and grapevines had coevolved to existed alongside each other. This cozy relationship hadn't developed with *Vitis*

Above Phylloxera leaf gall on an American vine (this is a rootstock variety grown in a conservatory in Spain). Many of the American vine species have coevolved with phylloxera and can live with it.

vinifera, and the arrangement was a hopelessly imbalanced one: the parasitism by phylloxera was too harsh, resulting in eventual death to the vine.

Phylloxera has a complicated life cycle, which was only worked out after the plague had broken. Like many aphids, phylloxera is parthenogenetic, which means it does not need sex to reproduce. The root-growing form settles on suitable roots and punctures them with its mouthparts. It injects saliva and this causes the root cells to grow into a structure known as a gall, which increases the supply of nutrients and provides some protection. Then it lays eggs. These hatch and move along the roots. Then they climb the trunk before being blown into the air. If they find a suitable feeding site, they will carry on proliferating and populations can increase rapidly.

In some regions there are asexual forms of phylloxera that can develop on the leaves. These also induce gall formation, and in this case the protective structure that encloses the feeding phylloxera extends below the leaf and opens to the upper leaf surface to allow crawlers out.

There are three potential mechanisms for vine damage by phylloxera: removal of photosynthates; physical disruption of the roots; and secondary fungal infections of damaged roots. It is unlikely that the severe, usually fatal vine damage is caused by the first mechanism. Much more likely is that the vine damage occurs through secondary infections by fungal pathogens.

GRAFTING—A CONTROVERSIAL SOLUTION

A controversial solution was proposed. American vine species, which harbored the invader in the first place, had natural protection against phylloxera. Where they had been planted they thrived, while all around them was a scene of devastation. Yes, the wine they produced tasted fairly bad but better odd wine than no wine at all, argued the faction, which became known as the *américainistes*. Tastings of wines made using the American vines were set up, but the depressing conclusion was that they were not good enough.

In 1869, a Monsieur Gaston Bazille suggested grafting vinifera varieties onto American rootstock. With the benefit of hindsight, this seems to be a brilliant solution, but at the time there were many unknowns. Chief among these were the following. First, how long would the graft last? Second, would the rootstock impart some of that American vine "foxiness" to the wine, altering the qualities of the grapes of the vinifera by this unnatural union of scion and stock? And third, how resistant would the various rootstocks be? It was already known that some American vines were more resistant to phylloxera than others. A period of intensive testing then ensued.

One of the first documented applications of this grafting was in 1874, when Henri Bouschet displayed an Aramon (*V. vinifera*) vine grafted onto American rootstock at a Congrès Viticole in Montpellier. It soon became clear that, somewhat counterintuitively, the wine made from vinifera vines grafted onto American rootstocks retained the full character of the vinifera variety while benefiting from the phylloxera resistance of the American roots. It completed a rather neat circle. The plague had come from America, but so had the salvation. The technique of grafting was easy enough to learn that just about anyone could attempt it. The supply of American vines was more of a problem, as many wine regions began to prohibit their importation to prevent the remaining phylloxera-free vineyards from succumbing to the pest. But not everyone was won over by this radical idea of grafting. Many resisted replanting, with the attendant three-year loss of production, and clung tenaciously to their chemical treatments. In the end, good sense prevailed, and the grafters won the day. The lengthy work of replanting France's vineyards began. It was not a straightforward process, and some of the grander estates, reluctant to pull out their vines kept them going with insecticide treatments as long as they could. The choice of appropriate grafting material was also complicated by the fact that it took a while to find American vine species that were well adapted to the chalky soils that predominated in some of France's key regions. One thing that replanting did facilitate was a change in the viticultural landscape, with some sites being abandoned and others being replanted with new varieties. In addition, the choice of rootstock, with its effect on graft parameters, such as vigor, became an additional variable, and a new item in the viticultural toolbox.

Grafting is an ancient practice that takes advantage of the fact that plants don't have an immune system, and thus different varieties or even species can unite and grow as one. It is recorded as far back as the 323 B.C.E. by Theoprastos for apple trees, and by Cato in *De Agri Cultura* for grapes in the 2nd century B.C.E. Physically, grafting involves the union of two plants, the scion and the stock, with the critical marriage being of a thin layer of cells known as the vascular cambium. This runs as a cylindrical layer of cells close to the outer circumference of the stem or trunk of the plant, producing phloem on the exterior side and xylem on the interior. Phloem is the vascular tissue that conducts sugars and nutrients, and xylem is the tissue that conducts water and dissolved minerals, as well as providing structural support.

Grafting takes advantage of the wound-healing response in plants. When an incision is made in a plant stem, the response is the growth of de-differentiated cells (cells that have reverted to an unspecialized state). Thus when branches or trunks that have already initiated secondary growth are damaged, they respond by growing callus, a tissue of undifferentiated cells produced by proliferation near the wound site. This same

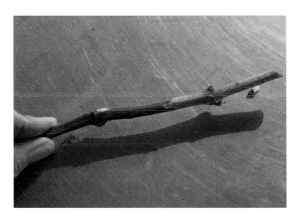

Above A grafted vine: this was bench grafted with an omega graft. Since the 1970s the omega graft has become almost universal in its use by nurseries, but now some are questioning how good it is: could it be part of the increased susceptibility to trunk diseases?

Above Grafted Arinto vines ready to plant in Portugal's Vinho Verde region. These have been grown in a nursery and are planted at the beginning of the growing season while they are still dormant.

callus production happens when the scion and the stock are joined together. Callus provides the tissue through which vascular continuity is restored, and signals from the vascular tissue of both scion and stock presumably influence the undifferentiated callus cells to become cambial cells. From these, the new vascular tissue, which is seamless between scion and stock, develops (Esau K., *Anatomy of seed plants*, John Wiley, New York 1977).

Grafting a vinifera vine onto an American vine stock is a relatively simple process. Incisions are made in both the rootstock and scion such that they marry together closely, giving some physical resilience to the join. The critical feature is to have the two cambial layers in apposition to each other. Typically, the cleft made in the rootstock is designed to allow the maximum area of cambial contact. Most grafting is done with cuttings (bench grafting), but it is also possible to graft in the field. And as well as joining American rootstock with vinifera scion, grafting is commonly used to transplant one vinifera variety to another in the field. In any graft situation, both scion and stock retain their own, separate genetic identity, and the graft merely facilitates vascular transportation. The advantage for the grower is they can switch the identity of their vineyards without the three-year loss of production that replanting would incur.

ARE WINES FROM UNGRAFTED VINES BETTER?

Have vines that have been altered by the grafting union lost something intrinsic to their identity? Were the wines made from ungrafted vines, pre-phylloxera, somehow better? "When I first met wine people it was the great debate," recalls Hugh Johnson, author and wine expert. "There was still lots of evidence of how good wines had been before phylloxera, but I don't remember a single conclusive demonstration that they had become less good since, or that if they had it was blamable on the louse." As the number of surviving pre-phylloxera bottles has diminished, the pre- and post-phylloxera debate has become less voluble. But when I wrote the first edition of this book, there were still people in the trade with enough experience of these pre-phylloxera bottles to have formed an opinion. Renowned wine critic and writer, the late Michael Broadbent was one. Are the pre-phylloxera wines better? "Who knows?" he replied. "The quality was undoubtedly high from 1844 until 1878: one has read about them and one has tasted them." A wine writer, and leading authority on wine, Serena Sutcliffe, is another who has experienced plenty of these old bottles. Are they special? "Yes, they are different. They are more intense; they have total, concentrated heart," says Sutcliffe. "They also have the most incredible scent and long, lingering finish. But you have to mix this in with the fact that yields were so much lower then: so how much is due to that, how much to the original vines? I suspect it is a combination."

Hugh Johnson also picks up on this theme. He agrees that pre-phylloxera wines may have been different, but this is not necessarily because of grafting onto American rootstocks. "Of course, it was not one event," maintains Johnson. "The oidium before it and the mildew at the same time made owners more proactive in their vineyards than they had ever been. They manured like mad to make their vines healthy and strong. They sulfured the vines, used carbon sulfate on the soil, invented Bordeaux mixture. Yields went up dramatically. They started *chaptalizing*, tried to control fermentation temperatures, started bottling much earlier, in fact, changed so many things in a short period that it's amazing that some of the 'pre-phylloxera' wines made during this time were as good as they were."

Sutcliffe adds that she doubts that the top wines from top vintages that are made now will last 80–100 years. "But how many people today require their wines to do that? Virtually no one. We are in the age of instant gratification, so it is perhaps irrelevant." Sutcliffe also adds a potential explanation for the longevity of the old wines. "A fascinating fact is that the vast majority of these pre-phylloxera gems are very low in alcohol, often ten percent or even under, as with some old Ausone from the 19th century: absolute proof that alcohol in itself does not assure longevity—perhaps the contrary, so beware all those fourteen percent numbers out there! It seems to eat up the wine over time."

As well as experience with pre-phylloxera wines, there are two further lines of argument that might give some insight as to whether ungrafted wines are somehow better, or not. The first is the fact that around the world there are some significant areas where vinifera vines are still grown on their own roots.

Perhaps most notable is southern Australia, which phylloxera hasn't reached yet, and which is kept clean by a strictly enforced quarantine. The Barossa in particular has a number of old-vine vineyards, planted on their own roots—these make some extremely fine wines, albeit in a rather different style to the Old World classic regions, which renders a comparison rather difficult. Most Chilean vines are also ungrafted, as are a large proportion of the vineyard area in Argentina. In Germany, the Mosel still has many old, ungrafted vines, with phylloxera struggling to grow in the impoverished slate soils of the region. The wines made from these old Riesling vines can be remarkable. The oldest vineyards in Central Otago are ungrafted (planted in the 1980s), and Oregon still has some older ungrafted vineyards (from the late 1960s). In both of these cases, phylloxera is present and their future is uncertain.

Useful evidence comes from particular isolated ungrafted vineyards still being cultivated in the middle of otherwise grafted wine regions. One example of an ungrafted vineyard where wines are made that can be compared with those from grafted vines on the same estate is Quinta do Noval's Nacional, a six-acre vineyard in the Cima Corgo of the Douro. Phylloxera has never affected this vineyard, which was first planted in 1925.

"There are some very old vines in the parcel, but I would say the average age is about forty years: when a vine dies we replant, but without grafting to any American rootstock," says Christian Seely, who took over running the Quinta on behalf of AXA in 1996. Each year, the grapes are picked, vinified, and matured separately. On average, just 250 cases are produced annually. "Although we don't declare it every year the Quinta do Noval Nacional is always extraordinary," says Seely. "The remarkable thing about the wine is that although we vinify it in exactly the same way as the other wines from Quinta do Noval, it is always very different from any of the other lots of wine from the estate. It marches to quite a different drum than the rest of the vineyard: sometimes it can produce a Vintage Port that is among the greatest wines of the world when the rest of the vineyard is having an LBV-quality year (as in 1996); at others it is not even of Vintage quality when the rest of the vineyard is making great Vintage Port (as in 1995), but it is always different to the rest of Noval. Even in years, such as the 1997, for example, when I believe the Quinta do Noval Vintage and the Nacional to be equivalent in quality, the wines have an entirely different character." Even here, though, we can't say that it is the fact that the Nacional vines are ungrafted, which makes them produce different wines. There may be something else about the soil of the Nacional vineyard that neuters the threat of phylloxera. This difference could reasonably be expected to influence the nature of the wine. As Seely puts it: "Nacional is a supreme example of the importance of terroir."

HOW DOES THE ROOTSTOCK AFFECT WINE-GRAPE QUALITY?

It is clear that rootstocks can influence the growth pattern of the scion quite considerably. Apples are instructive here. In the early 1900s scientists at the East Malling Research Institute in Kent in the UK did a lot of work on rootstocks for apple trees. There are around 20 well-known rootstocks for apples, and these determine how the apple tree (the scion) will grow. With one rootstock, a certain variety of apple will produce a tall tree. With another, the same variety will produce a dwarf plant just 1.5m (5ft) high. Not only does the rootstock provide a supply of water and mineral nutrients for the plant, it also communicates with the scion by means of hormonal signals. That the roots of grapevines signal to the aerial part of the plant is well known, particularly from work done recently on partial deficit irrigation and precision root drying. The roots are able to signal information about soil water status to the aerial parts of the plant by means of the stress hormone abscisic acid, which is produced in response to deficit. This results in the leaves closing their stomata to avoid losing water before they are themselves suffering any water stress. It is likely that the rootstock will be signaling all manner of information by means of plant hormones, and there will be communication the other way also, from shoots to roots.

While the scion and rootstock having subtly different genetic makeup, the rootstock choice is likely to have physiological consequences for the vinifera scion and will, to a degree, shape its growth pattern. This interplay is effectively a new viticultural tool, provided that viticulturists understand enough about it to manipulate it effectively. This will in turn have an impact on grape, and thus wine, quality. But there is no reason why this should be a negative impact. In many instances it could be positive. Science has no evidence to support the claim that grafted vines produce grapes that are necessarily inferior to those produced by the same variety grown on its own roots.

The fact that so many expert commentators over the centuries have argued that there was something special about the pre-phylloxera wines is strongly suggestive that they were very good wines indeed, and perhaps even better than the wines that followed, made from grafted vines.

There are numerous confounding factors here, not least that old-vine vineyards were being uprooted at more or less the same time and being planted with new vines. While no one knows for sure why it is the case, young and adolescent vines simply don't perform as well as their older siblings. It could be this dip in quality following widespread replanting that caused commentators to note a qualitative deficit post-phylloxera, which may then have been wrongly attributed to grafting.

OTHER PESTS

PIERCE'S DISEASE

This has been a major problem in many wine regions in North America. It is caused by a bacteria, *Xylella fastidiosa*, which is introduced to grapevines via the mouthparts of its vector, the glassy winged sharpshooter. The bacteria grows in the xylem and blocks it, killing the vine within a few years of infection. It was first reported in the mid-19th century when thousands of acres of vines in the Anaheim region near Los Angeles were wiped out; PdR1, the first Pierce's disease resistance gene was discovered in 2006. Marker-assisted selection has been used to breed this gene into a *Vitis vinifera* background, and five new resistant varieties were released to nurseries in the USA in 2017.

FLAVASCENCE DORÉE

This is a big problem in European vineyards. It first appeared in the 1950s, with the introduction of a leaf-hopper called *Scaphoideus titans*. This is an insect vector for a phytoplasma called *Flavascence dorée phytoplasma* (FDp), a sort of bacteria that lives in the guts of *S. titans* and also the sieve plates in the phloem of the grapevine, which *S. titans* feeds on. FDp belongs to the Elm Yellows group of phytoplasmas, and there are five similar organisms that also affect grapevines. Together this problem is called grapevine yellows. This is because the leaves of affected vines go yellow (or red, in some cases), and curl downwards. The flowers wither, berries shrivel, and the canes don't ripen properly. The vines decline and eventually die. Such is the extent of the problem that some regions insist on compulsory insecticide treatments when FDp becomes an issue—even for organic producers. They do this because the only way to deal with *S. titans* is to wipe it out over a large area, because its range is quite extensive.

DROSOPHILA SUZUKII

The common fruit fly, *Drosophila melanogaster*, is a regular feature around wineries at harvest. These tiny flies seem to get everywhere. They are annoying, but not a major problem, although they can spread acetic acid bacteria. And if you get a fly in your wine, whether or not it is going to alter the taste depends on the sex of the fly: male flies are okay, but the females release a pheromone that can taint the wine.

In recent years, however, there's been a new fruit fly on the scene, and this one is much more of a problem. It's *Drosophila suzukii*, also known as the spotted wing Drosophila, or the Asian fruit fly (its origins are from Japan). It arrived in Europe and California in 2008, but only came onto the radar of wine producers in 2014, when it was a problem in both Burgundy and Champagne.

Why is it bad? It's because it has a serrated ovipositor. This is the device that females use to lay eggs. The regular fruit fly cannot penetrate grape berries, and so can only lay eggs in already-damaged fruit. *D. suzukii* can pop an egg through the berry skin, and then larvae can develop inside the berry, with little evidence that anything has happened until too late. The result is massive problems with sour rot, and the problem is especially acute with red grapes. Females can lay up to 21 eggs a day and 195 during their lifespan, and this fly can manage 13 generations a year, so it can really become a major problem fast. There are no simple solutions. Insecticides work, but have to be reapplied often, and are extremely unfriendly to the environment. There

Above Sexual confusion being used in Bordeaux. The small brown capsule seen hanging on a wire among the vines contains pheromones for the grape berry moth, which disrupt its ability to find mates.

are currently no biological remedies, and perhaps the best solution is to spread traps throughout the vineyard to lure the flies away from the grapes.

GRAPE BERRY MOTHS

Moths are a common problem in vineyards. There's the European grapevine moth (*Lobesia botrana*), the American berry moth (*Polychrosis viteana*), and the Australian light-brown apple moth (*Epiphyas postvittana*). All do the same thing, laying eggs on the vine and then the larvae feed on flowers and fruit, with an additional problem being secondary infections of grapes that they damage. This is an insect problem where there is an elegant solution: sexual confusion. You might have seen small plastic pods hanging on wires in vineyards among the leaves. These contain the pheromone that males use to locate females. When the vineyard is full of this, then the males can no longer find their mates, hence the term sexual confusion. It is more expensive than insecticidal control, and it only works if enough people in the same area use it. But it is a great example of using biology to solve problems.

MEALYBUGS

Named after the powdery secretions on their skin, these sap-sucking insects are common in vineyards. One of the reasons they are a problem is that they are vectors for leafroll virus, spreading it as they move from vine to vine.

LEAFROLL VIRUS

Do you love those autumnal colors in the vineyard, when it becomes a mass of red, orange, and yellow at the end of the growing season? To a viticulturist, it's not such a happy picture, because it is often caused by leafroll virus. The main viral concern for viticulturists is Grapevine leafroll-associated virus type 3 (GLRaV-3a). Affected vines struggle to finish the ripening cycle. For white wines, this problem can be spotted by looking for leaves with rims that are rolled down (although some varieties don't show this). For reds, it is much more apparent with bright red colors developing in the leaves at the end of the season (the veins remain green), as well as the rolling of the leaves.

It has been a particular problem in South Africa but is present everywhere. J.C. Martin of Creation Wines in South Africa's Hemel-en-Aarde region wanted virus-free vineyards so he decided to plant his vineyard on virgin ground, allowing him to start the vineyard virus free, back in 2002. "At that time virus-free vines were first available from VitiTec," he recalls. "It has been 15 years now and there is no virus on this farm." Martin has convinced neighbors La Vierge and Domaine des Dieux to be equally vigilant when it comes to leafroll virus, which is common in South Africa's winelands.

Although some have argued that virused vines can produce excellent wine, in part because they delay ripening, the general view is that it is detrimental to quality and is best avoided. "From a quality point of view you don't want to pull the vineyards after 15 years," says Martin, which is what often happens when there's lots of virus around. The vineyards struggle to ripen reds and gradually become less productive. And then, when you have pulled out the virus-ridden vineyard, you need to leave the soils fallow for

five years because they are contaminated, or else the whole cycle will start again (the virus can spread through mealybugs and also soil nematodes). It carries with it a huge economic impact, which is why many producers simply haven't been able to tackle their virus problem.

The strategy for dealing with mealybugs at biodynamic winery Waterkloof has been smart. To begin with, they increased the biodiversity on the farm. Then they brought back wasps and ladybugs, which are the natural enemies of mealybugs. Initially, mealybug levels had been seen at 800–900 per block. This was brought down to 40 or 50. Now they are even fewer, and it is no longer a problem. The levels are monitored by pheromone traps, and the highest levels are at the boundary with neighboring farms, around 30 per block.

In New Zealand, virus is a concern. "Virus is becoming a bigger and bigger issue," says Steve Smith of Smith & Sheth. "It's the greatest threat to New Zealand making wines of old vines, especially for red wines." While the virus has likely been in New Zealand since the 19th century, it became a particular problem only recently. "Rapid spread began with the boom in vineyard plantings starting in the 1990s," says Dr. Simon Hooker, general manager for research at New Zealand

Above Leafroll virus. This is a Chardonnay vineyard in Franschhoek, South Africa. With white varieties the leaves don't take on the red coloration that they do with red varieties. This picture was taken shortly after vintage, and toward the end of the growing season the leaves start to roll like this.

Wine Growers. "The disease was noted as causing declines in fruit and vine quality by the early 2000s, especially with regard to the production of premium red wine styles." In response, New Zealand Winegrowers began working with the nurseries to develop the guidelines that would eventually become the Grafted Grapevine Standard, and in 2008 began the Virus Elimination Project. He says that the project has already achieved considerable success in the first two pilot regions in Martinborough and Hawke's Bay. "Many growers in those regions have already achieved the initial goal of reducing infection incidence to below one percent in their vineyards—something they would not have believed possible five years ago."

LADYBUGS

Ladybugs aren't a problem in the vineyard. It's just when they end up on grape clusters and then find their way into the wine that they are an issue. When stressed, ladybugs release high concentrations of methoxypyrazines, which can end up tainting the wine. For this reason, winemakers working in an area where this is a risk use vibrating sorting tables to get rid of any ladybugs on clusters. Vineyards near soybean farms can be at risk because once the beans have been harvested, the ladybugs head off in search of a new home. If this happens to be a vineyard, then they can be present in significant numbers during harvest.

TRUNK DISEASES

There's nothing that gets the heart of a wine geek beating fast like the sight of an old vineyard with its stumpy, gnarly old vine trunks that have been in the ground longer than most people have been alive. Old vines are prized the world over, but there is a problem. They are at risk. Visit an old vineyard and you'll see lots of gaps. Increasingly, vines are dying before their time, rotting from the inside. The cause is a range of fungal diseases known collectively as grapevine trunk diseases (GTDs), affecting the woody parts of vines, eventually causing vine death. They are on the rise at the moment, and celebrated viticulturist Richard Smart likens them to phylloxera in terms of their potential impact on winegrowing. Why are they on the increase? And what can be done to save old vineyards?

There is nothing new about trunk disease. "There has been lots of talk about trunk diseases," says Dr. Mark Sosnowski, one of the world's leading experts on this topic, from SARDI in Australia, "but these are as old as viticulture itself." They are pretty complex. Unlike powdery and downy mildew, which are each caused by one species, these diseases are often caused by a number of different fungal species. With the mildews, scientists know what causes them, how the disease progresses, and how to manage them. Not so with the GTDs. An extra factor that makes them complicated is that some of the fungi are endophytes that actually live in the plant tissue without causing any harm, and then suddenly switch to being pathogens. They may be present in the vineyard but not causing any symptoms of disease. It could be that some of these problems are initiated by vine stress: then, the previously harmless fungi in the tissue become a problem.

José Ramon Úrbez-Torres, based at the Summerland Research and Development Centre in Canada's Okanagan Valley, is one of the leading experts on GTDs, and he says that more than 130 species from 34 genera are known to be involved in causing these diseases. The term "grapevine trunk disease" is a fairly new one and was first proposed in the late 1990s by Luigi Chiarappa.

Even though these diseases are thought to be as old as grapevine cultivation itself, they are currently in the spotlight, and the evidence is that they have been on the increase in the last two decades. Why? Úrbez-Torres suggests three reasons. First, since the early 2000s some of the previously effective chemical controls, such as sodium arsenate and benomyl, have been banned. There may also have been some changes in viticultural practice, with an increasing move to systems with more pruning wounds or bigger pruning wounds. And although there has been a general worldwide reduction in vineyard area, there has been a boom in planting/replanting in some areas, and this has put pressure on nurseries. The result? Vineyards have been established with poor-quality planting material.

The earliest scientific studies were undertaken on the trunk disease ESCA, in France, at the end of the 19th century. Initially it was called *folletage*, which translates as "apoplexy." Ravaz, the famous viticulturist, noted vines suddenly withering and then dying. He made the connection between this and a very similar disease that had been described in Smyrna, in Turkey, called Iska. Another botanist, Pierre Viala, found out from a friend that in Italy, a similar disease was known as ESCA. Viala proposed that this name should be adopted, and it stuck. Vines are a potential food source for fungi because, like most woody plants, they have carbohydrates stored in them. Because vineyards are pruned every winter and early spring, the cuts open a way into the tissues. However, most of the trunk diseases that are worrying viticulturists seem to attack older vines. Examples would be ESCA

complex, and the dieback diseases Botryosphaeria and Phomopsis. But there is a second group of trunk diseases affecting young vineyards such as black-foot and Petri disease. In this second group, the first signs are often a loss of vigor, where vines look like they are struggling. If you cut into the wood, then you can see some discoloration.

CONSIDERATIONS FOR TRUNK DISEASES

In 2019, I was in Crete visiting the vineyards of Lyrarakis. Every now and then, I'd spot a vine with the head of its trunk split down the middle, and a stone inserted in the gap to keep it from closing. This is an ancient remedy for trunk disease, intended to dry out affected wood and stop any fungal infection in its tracks. This has clearly been an issue for a long time. But my first real encounter with trunk disease in the wild came in the Loire Valley of France, in the vineyards of Sancerre. It was April, and the vines had just started growing again, a hopeful time of year, but also a nervous one because of the risk of frost. The soils looked amazing: so much limestone, of a few different kinds, and in some of the plots there was flint, too. These were proper winegrowing soils. But there were lots of gaps in the rows, especially in the older plots. These venerable old vines seemed to be dying

Above A stone placed in a cleft in a vine trunk in Crete, in Greece. This is an ancient way of dealing with the early stages of trunk disease.

Above A vine trunk that has been sawn across shows evidence of the disease inside.

in significant numbers. Of course, you can replant, but it is hard to get young vines established in the middle of an existing row, and they will take many years to fall in step with their older neighbors.

Grapevine trunk disease is the culprit, and especially in this region, ESCA. Some varieties are more susceptible, and Sauvignon Blanc is one of them, which is why the Centre Loire has been a hotspot. Chenin Blanc, Cabernet Sauvignon, Ugni Blanc, Grenache, and Syrah are other varieties that are thought to be susceptible, while Sémillon, Merlot, and Aligoté are less affected. The scope of the problem is astonishing: in the Loire it is thought that more than 20 percent of vines are affected, while in France's entire vineyard the incidence is estimated to be 13 percent.

"In the region, when we still had sodium arsenate, no one talked about ESCA or BDA (black dead arm), but there was Eutypa," says Benoît Roumet, of the Vins Centre-Loire, a trade body for the region. He says that Eutypa, another sort of trunk disease, was affecting Sauvignon Blanc, but was not harmful to other grapes and was not a huge problem.

But in 2001 arsenate was banned, and suddenly ESCA and BDA emerged as big problems. We spent some time in the vineyard and I asked Roumet what was being done about trunk disease. He pointed out some of the vines that had strange alterations to their trunks through the vine equivalent of dentistry using a chainsaw. This process is called *curetage* and is one of the ways that growers deal with this issue.

François Dal, a researcher who works with Sicavac in the Centre Loire, has been very important in the fight against trunk diseases. He has written a book that is a practical guide for winegrowers about how to deal with this problem.

Roumet explained how their thinking about trunk disease had changed over time. The current thinking is that dead wood in the vine might actually be what is causing it to die. "Because we have dead wood, the fungus can come into the vine and make it weaker," he says. "People have taken vines that seem healthy and found lots of dead wood. This made people rethink how they are pruning."

Dal has been an advocate of an old, almost forgotten way of pruning called Guyot Poussard, which was first named after its inventor back in the 1920s. Dal has been championing this technique as a way to reduce trunk disease, and it has been made famous through the work of Marco Simonit of pruning consultancy Simonit & Sirch, based in Italy but with a network of experts spread across the wine world. The main idea behind this pruning system is protecting the sap route in the vine, which means taking new canes from the outside each time, and so there are pruning wounds only on the upper side of the main vine structure. There is also a version of this approach for vines with a permanent cordon and spur pruning. In addition, pruning cuts are taken some distance away from the main trunk to avoid what is known as the "cone of desiccation,": here, there is a little dieback because the vine is not good at healing wounds, and if the cut is made too close to the permanent structure, this cone of dead wood impedes sap flow. The result of the Simonit/Dal pruning techniques is fewer and smaller pruning wounds and less impact on the sap flow. Ideally, Guyot Poussard should be practiced on a new vine from the outset, in which case it takes an extra year to get to production, but this sacrifice is worth it from the extended life of the vine. Simonit has published two books on pruning—one for cane pruning and the other for spur pruning—and these have now been translated into English. His consulting team is in high demand worldwide.

HOW TO DEAL WITH TRUNK DISEASES

There are two distinct types of trunk disease. For young vines, up to six years old, there's black-foot and Petri disease, together referred to as "young vineyard decline." Then there are the dieback diseases affecting more mature vines: Eutypa,

Above An old vine shows the scars from the chainsaw work that was used to remove any ESCA-infected wood using the *curetage* technique.

Botryosphaeria, Phomopsis, and ESCA. These are thought to be caused by fungi colonizing wound sites, typically resulting from pruning cuts. Each of these trunk diseases is caused by a range of species of fungi, and now sophisticated DNA techniques exist that can identify them very accurately.

The economic impact of GTDs is enormous. A 2001 study put the cost of trunk diseases in California alone at $260 million a year. In Australia, the yield loss in Shiraz from trunk diseases is estimated as A$20 million (approximately $14 million USD) a year. Globally, the cost of replanting just one percent of the vines worldwide is $1.5 billion.

In southeastern Australia, surveys have picked up dieback signs in vineyards from as early as five years after planting. By 15–20 years the incidence of at least some of the symptoms in some vineyards is 100 percent. New Zealand surveys have shown that Sauvignon Blanc is rather susceptible, with incidences of 60–80 percent in 15–20-year-old vineyards.

There is lower risk the closer pruning is done to spring. Rain is also a factor: 2mm (0.08in) of rainfall is enough for the fruiting bodies of the fungi responsible to spore, and the more the rainfall the more the risk. In California, some vineyards do double pruning: a mechanical first pass, then finishing by hand later on, which reduces risk. It's also possible to protect pruning wounds. Acrylic paint works well, but there are also products that do a similar job, such as Greenseal, Garrison, Gelseal, and Baseal. There exist fungicides for spray application, such as Fluazinam. There's also biocontrol, such as Vinevax (which is the fungus Trichoderma). This can be an excellent protector, but reliability is an issue. Garlic and lactoferrin (a by-product from milk and cheese production) have also been used.

Finally, there's remedial surgery. In this case, you have to chop the trunk down to the point where the wedge of dead material is no longer seen, and then grow the vine back up. If this is done well, it works, because the established root system helps the vine grow back relatively quickly..

At Familia Deicas in Uruguay they regularly renew vineyards like this every 12 years, by growing up a second trunk that is then used to replace the older one. One year's production is lost, but the vines are healthy, grow consistently, and have an old, established root system.

Both Simonit and Dal are promoting a different sort of remedial surgery, called *curetage*. Here, they use small chainsaws to remove any soft ESCA-infected wood. The idea is that if they spot a vine showing the first symptoms of decline—the "tiger stripes" on the leaves— the vine can recover if they remove the infected wood producing the toxins. It is hard work, and time-consuming, but justifiable if the vines are old and the terroir is a good one, making high-priced wine.

Charles Lachaux of Domaine Arnaud-Lachaux in Burgundy has been combating trunk diseases by using the Guyot Poussard pruning technique as well as doing *curetage*. The result is no more ESCA, better yields, and healthy vines. But it takes time. For his Suchots plot of half a hectare it took 36 days (he worked alone), and for his one-third hectare holding in Romanée-St-Vivant it took 20 days. But for old vines in famous vineyards, it is worth the effort.

7 Vine immunity and breeding for resistance

GRAPEVINE IMMUNITY

We talk of humans being immune, and we talk about plants, such as the grapevine, being immune, but we are not really comparing like with like. It is the nature of the infection and the method of response that differs.

Both plants and humans share two ways of defending themselves. The first, which is common to both, is what is called innate immunity. These are the built-in, nonspecific defenses, structured to fight off attack in a more generalized way. It begins with physical barriers: humans have skin, while plants have a waxy cuticle and other defenses such as hairs and spikes.

But both humans and plants need to exchange gases with the atmosphere, and this creates a potential route for bugs to gain entrance. Humans have the huge, warm, moist, more-or-less unprotected surface area of the lung, while plants have stomata, small openings to the outside, scattered on their surface. And some pathogens and insects have found ways of getting through the physical barriers of skin and cuticles.

Plants are also chemical factories and produce a wide range of "secondary metabolites." Some of these are used to dissuade animals from eating them, such as tannins. This is because tannins bind to proteins, including the digestive enzymes of herbivores, thus making them unpalatable.

In this chapter we will explore the nature of grapevine resistance to disease, which we call immunity, and then look at how this can be enhanced. There are real limits to this immunity in *Vitis vinifera*, and so then we will explore attempts to breed new resistant varieties.

STAGE 1: PAMP-TRIGGERED IMMUNITY

Aside from any physical barriers, the most important part of innate immunity relies on recognizing chemical signals common to the sort of pathogens that might be a threat. As mammals, we have several different types of circulating white-blood cells tasked with recognizing these kinds of danger signals and responding appropriately. Plants don't have the equivalent circulating cells, but instead, as a first line of defense, have pattern recognition receptors (PRRs) on the outside of their cell membranes. These recognize specific molecules associated with pathogen or herbivore attack—both from microbes themselves, and also from their own tissues that have been damaged. The signal from these receptors triggers a complex metabolic response that, hopefully, stops further colonization of the plant tissue. This is called PAMP-triggered immunity (PTI, where PAMP stands for pathogen-associated molecular pattern). PAMPs can also be divided into two groups. The first is molecules associated with pathogens directly, called microbe-associated molecular patterns (MAMPs), and the second is molecules related to the plant itself, when it is damaged, called host-derived damage-associated molecular patterns (DAMPs).

Perception of PAMPs by the plant switches on genes involves in signaling in the plant cell, which in turn leads to the synthesis of wide range of defense molecules. Pathogenesis-related (PR) proteins are important here and include enzymes that attack the pathogen cell walls. Antimicrobial compounds such as phytoalexins are also switched on (resveratrol is a famous phytoalexin that has been widely studied for other reasons). The cell wall of the plant is also reinforced. Salicylic acid, jasmonic acid, and ethylene communicate news of the attack to other parts of the same plant and sometimes also to neighboring plants, allowing them to have defenses in place should the pathogens attack. This basal level of immunity bolsters the plant defenses but isn't always enough to stop further infection. This is when stage two kicks in.

In humans, stage two is adaptive immunity. Humans also have an adaptive immune system, where specialized white blood cells are able to learn to recognize self from non-self, create an immune memory, and then urge other white blood cells to respond to specific threats. So, when someone gets sick, the adaptive immune system recognizes the bacteria or virus as non-self and begins to generate antibodies against it. All being well, the resulting defense response targets the bug specifically then eliminates it from the body. Afterwards it keeps some memory of this attack and this will prevent a subsequent infection. This system is the basis of vaccination. By injecting an inactivated bug, or fragment of a bug—along with an alarm signal called an adjuvant—medics are able to prime the adaptive immune system to make people resistant to infection by the real thing. The adaptive immune system is a remarkable, complex system, and its utility is highlighted by what happens when it goes wrong or is compromised. Plants have no such similar system.

STAGE 2: R GENES

Plants do have resistance genes, though, which are the next line of defense. Some pathogens can produce molecules called "effectors" that interfere in this initial PTI process, leaving the way free for them to make use of the plant's resources unhindered. This is when the second stage of plant "immunity" kicks in. Plants have a set of genes called R genes that produce members of a family of nucleotide-binding leucine-rich repeat (NLR) receptors. When they are activated by the effectors

released by the pathogen, a process called effector-triggered immunity (ETI) kicks in. This results in a robust response that usually results in the death of the cell. Called the hypersensitive response, this sounds like bad news for the plant, but this cell death stops the pathogen from spreading. Plants typically have around 1,000 different R genes, which allows them to recognize 1,000 different pathogen avirulence (Avr) genes. The avirulence genes are pathogen genes that code for the effectors that interfere in the PTI process. But if plants lack the right R gene, then they cannot defend themselves against a specific pathogen in this way. Plants have no adaptive immunity, so it is not possible to improve the resistance of your grapevines to microbes that they have no R gene against, by "boosting their immunity" in some way, other than by making their generalized responses (stage one) more effective.

STEP 3: THE ARMS RACE

But even if a plant has the right R gene, the arms race can carry on. Through evolution, skilled pathogens can avoid ETI by developing new effector genes that the plant lacks an R gene for, or by acquiring new molecules that can suppress ETI. In this case, the pathogen will win the battle. This is an important fact: just because a plant is resistant now, it won't always be. Any resistance strategy that relies on just one R gene is going to be fragile.

STAGE 4: RAISING THE ALARM

This sort of encounter between plant and pathogen can result in long-distance signals being sent to alert other parts of the plant or even neighboring plants to get ready. This process involves two pathways that are given the confusingly similar names systemic acquired resistance (SAR) and induced systemic resistance (ISR). These involve systemic (sent through the whole plant) signals that prime the defenses, so they are ready to go. SAR involves an unknown signal, acting together with salicylic acid, and is triggered by herbivory or disease. ISR is triggered by infection, also by reactions of roots with rhizobacteria, and the signaling is mediated by the plant hormones jasmonic acid and ethylene. While direct induction of host defenses has a high metabolic cost, priming the defense responses so they are ready to go has only a small cost.

MAKING GRAPEVINES MORE RESISTANT

There are four main pathogens affecting grapevines. *Botrytis cinerea*, *Erysiphe necator* (powdery mildew), *Plasmopara viticola* (downy mildew), and then the several different fungi that cause grapevine trunk diseases (see chapter 6, page 89). It's powdery mildew and downy mildew that result in much of the spraying that takes place in vineyards worldwide, and that is because *Vitis vinifera* doesn't have resistance to these American imports: it lacks the right R genes.

Wild *Vitis* species in the USA and Asia do have resistance genes, though. For downy mildew resistance, some 27 genes have been identified as potentially involved, and resistant varieties have been developed by breeding *Vitis vinifera* varieties with American species. Resistant vines that have been bred such as Regent and Cabernet Blanc have a gene called Rpv3 in them, which is an R gene that gives them resistance to downy mildew. Resistance to powdery mildew in hybrid vines has been shown to be due the R genes Run and Ren that have come from their American wild *Vitis* ancestry.

"The best way to obtain resistance against downy or powdery mildew is to identify then combine at least two R genes from resistant Asian or American species in the *Vitis vinifera* genome," says Benoit Poinssot, a researcher from the Université de Bourgogne in Dijon. The combination of R genes is known as pyramiding and Poinssot says that it is important to avoid the emergence of new effectors encoded by mutated Avr genes of the pathogens under selection pressure. And when it comes to powdery mildew and downy mildew, "when the natural genome does not contain

an R gene, the grapevine will be susceptible," he says, "and that's the case for *Vitis vinifera*."

So, what can be done to aid immunity in vinifera, which lacks the right R genes? "We can trigger plant immunity by elicitor molecules," says Poinssot. Elicitors that can induce SAR enhance the production of phytoalexins (defense molecules). The main phytoalexins in vines, trans-resveratrol and viniferin, are fungitoxic against *Botrytis cinerea* and enhance resistance to *P. viticola*.

How do elicitors work? They mimic the attack of the grapevine by pathogens. Normally, vinifera fails to recognize the danger posed by downy mildew and does not mount sufficient defenses. So, inducers are danger signals that the vine does recognize, setting off the appropriate responses that can help protect the vine. These elicitors include chitosan, laminarin, oligolacturonate, b-aminobutryic acid (BABA), and benzothiadiazole-7-carbothioic acid S-methyl ester (BTH). There are also microbes that can be used, such as *Trichoderma harzianum* strain T39, which works by activating grapevine resistance locally and also systemically.

These trigger PTI in the plants. "This works very well in greenhouses (we did many articles about that), but the results are more variable outside in vineyards when we do scientific experiments." There are many hypotheses about why the promising results in greenhouses don't translate to the vineyard. He mentions plant physiology, climate, and natural interactions with soil microbes. "We are studying these hypotheses."

The frustrating thing is that stimulating the immunity of vines works, and the science behind it is good, but it has not yet worked in the vineyard to the extent that it could be useful. "Stimulating the plant immunity is scientifically demonstrated and it can give very interesting results in control conditions," says Poinssot, "but actually in vineyards, the best way to avoid chemicals is to introduce new resistant cultivars that integrate at least two R genes against *P. viticola* and two others against *U. necator*. In this case, experiments realized in France have shown a reduction of eighty percent of the chemical uses compared to twenty–fifty percent with biocontrol products."

Aziz Aziz is a researcher at the Université de Reims, and I quizzed him on the potential for inducing resistance in vines, for example through the use of biodynamic preparations. "Biodynamic preparations, like other biocontrol approaches, such as beneficial microbes, natural or bio-based molecules, microbial or plant extracts, are likely to prime plant immunity," he says. "These microbes or compounds are generally perceived by the plant through different mechanisms and the plant immune reaction can be boosted locally and systemically. These reactions result in a strong and rapid expression (priming principle) of effective defenses (often nonspecific) against various diseases like powdery and downy mildews. *Vitis vinifera* can therefore become less susceptible to these diseases if it is potentiated at the right time and at a sufficient level to express faster and stronger defense potential."

BREEDING RESISTANT VINES

The fact that *V. vinifera* lacks the R genes that would confer resistance to downy and powdery mildew has sparked an interest in breeding resistant vines that do not need spraying. But perhaps there has not been as much interest in breeding programs as there should be. This is because it is not possible to keep the same varieties that we know and love and do any breeding, unless genetic modification is used—and this is considered unacceptable in the wine world. If we want a more resistant Pinot Noir, all we can do is hope that some mutation emerges naturally that confers a bit of resistance, and then clone that vine. This is because when vines have sex and produce a new vine from seed, it will be a new variety.

In the eastern USA, since the 17th century there have been repeated attempts to grow *V. vinifera* and all failed because of disease problems. Native American vines had some innate resistance to powdery and downy mildew, which were the main reasons that vinifera failed here. Of course, phylloxera would also be a big problem, but the vines usually didn't survive long enough for this to take them out. The earliest reports of deliberate attempts to breed resistant vines date back to the 19th century in the USA. There are reports from 1822 of Harvard researchers breeding hybrids between the European vines and indigenous grapes, but things really kicked in by the middle of the century when breeders such as Thomas Volney Munson, Hermann Jaegger, William W. Valk, and Nicholas Herbermont started considerable breeding experiments. These programs produced what are known today as the American hybrids. The goal was to produce varieties that could be grown in the problematic eastern USA where vinifera failed to flourish, combining the perceived superior wine quality of the European vines with the hardiness of the native American varieties.

Later that century there was a renewed interest in breeding not only disease-resistant varieties, but also phylloxera-resistant rootstock and vines, as the phylloxera crisis hit hard across the world. These are known as the French hybrids. While the rootstocks bred then are still being used today, the interest in growing hybrid direct producers (known in France as *hybrides producteurs directs*, HPDs) waned somewhat when the option of grafting onto resistant rootstock became the solution of choice.

More recently, there has been renewed interest in breeding resistant varieties as attention has shifted to the sustainability of vineyards. Increased knowledge of genetics has given this initiative added impetus, as we now know the resistance genes from native American and Asian varieties that are needed in crosses in order to confer significant resistance to downy and powdery mildew.

How is vine breeding done? First, you need to decide which vine is going to be the mother and which will be the father. The native vines are usually dioecious (with separate male and female vines), but almost all cultivated vines are monoecious (hermaphrodites, with both male and female organs on the same flower). So, step one is emasculation of the mother variety, a fiddly process that involves taking the anthers of each vine flower with tiny tweezers to leave just the ovule and pistil. This needs to be done before pollination occurs. Then this flower cluster is bagged to prevent any random fertilization. The next step is to collect the pollen from the male parent: for timing reasons, it is best to make the earliest flowering variety the male, because you can keep pollen for a while, but the fertilization needs to take place at exactly the right time. This is when you can see the stigmatic fluid appearing in the stigma of the pistils (the tip of each flower). All it needs is for some pollen to germinate in this fluid. The pollen produces a pollen tube that grows toward the ovule and fertilization happens. One hundred days or so later and you have a grape with mature seeds in it, which you collect and sow.

The result of a breeding program like this is lots of seedlings, which will grow into grapevines. Each will be different, and so the long process of examining them for positive features begins. Most won't be any good: you are looking for a combination of organoleptic qualities (you want a good yield of grapes that have the potential to make tasty wine) and also resistance. This all takes a long time. It will be three years until you get grapes, and then there is a cycle of taking cuttings of the most promising candidates and then growing them to go through further rounds of selection. From the crossing to a new variety using this conventional breeding approach, it might take 25–30 years, and the vast majority of the crossings are relative failures.

The process can be sped up by a modern technique called marker-assisted selection. This relies on what are known as quantitative trait loci (QTLs) of important traits. These are like genetic bookmarks: it allows you to see whether your crossing has the genes that you are looking for. This allows for optimization of the crossing combinations, and it makes identifying promising offspring much quicker. Marker-assisted selection speeds up breeding because it's possible to see whether or not the desirable resistance genes are present almost immediately, rather than having to grow the vine for a season or two and test it in experiments. Using marker-assisted selection, scientists can find traits of interest in wild vines, and then by a series of crossings and back-crossings the vinifera percentage in the offspring can be ramped up higher, retaining the desirable character from the wild vine. The modern resistant varieties can therefore have vinifera genome content of 90 percent—high enough that they are being included in the national catalogues of various countries as *V. vinifera*, rather than hybrids.

For the future, the goal for breeding would be to identify markers of good fruit and wine composition, which would speed up the process dramatically. At the moment, we can't define wine quality in molecular terms. In order to see whether a new variety has the potential for making high-quality wine, we have to wait for grapes and then make some wine. This holds up breeding programs. A set of quality markers that could be screened for with marker-assisted selection would make things much quicker.

PIWIS

On a recent trip to Germany's Mosel wine region, I drove up the slopes behind the village of Kröv. While the Mosel is famed for producing some of the world's top Rieslings, this was not the focus of the trip. I'm here with Jan Matthias Klein, who has plans to plant a vineyard with a difference, in a place called Kröv Paradies. Rather than Riesling, the mainstay of the vineyards around here, he is going to plant two hectares with PIWIs over the next couple of years. He showed me the plot where the schist soils are already under preparation.

What are PIWIs? The name stands for Pilzwiderstandsfähige, which are specially bred,

fungus-resistant grape varieties with at least 85 percent vinifera in their genomes. These are technically hybrids, but the people behind PIWIs don't like you to refer to them that way. PIWI International is the organization responsible for promoting these varieties. Founded in 1999, it now has more than 550 members from 21 different countries in Europe and the USA.

Klein makes two lines of wines in his winery, Staffelter Hof. The first is a more classical range, from organically grown grapes, focusing on Riesling. And then there's his natural wines: these have cartoonlike labels and are made without any sulfite additions. But even though he farms organically he, like everyone else here, has to spray far more times than he would like.

The PIWIs could be a solution: they are bred for resistance to the fungal diseases that cause so many problems for vinifera. If you work naturally in the winery, why not work more naturally in the vineyard, too? These new varieties are doing well in blind tastings. "Quality-wise they are on a par with traditional varieties," says Klein. "Maybe not Riesling," he adds with a smile. The varieties he will plant are Muscaris, Souvignier Gris, Sauvetage, Sauvignac, and Donauriesling.

Not far from Kröv is Traben-Trarbach, where Markus Boor of Weingut Louis Klein is also taking an interest in PIWIs. His focus is not on making natural wines, though. About ten percent of his production is PIWIs.

"It is the future," he says. "For anything other than Riesling we have to plant PIWIs." A tasting through the cellar showed that there is nothing second-rate about these PIWIs. Typically, he says he will spray these vines 2–3 times a year, whereas the Riesling needs spraying 8–12 times. "We have had years without spraying the PIWIs," Boor says, although it is necessary to spray occasionally just so they keep their resistance. His varieties include Sauvignac, Donauriesling, Cabernet Blanc, and Johanniter.

RESDUR

France has a resistance breeding program called ResDur. The roots of this program go back to the 1970s, when Alain Bronner of INRA in Montpellier identified genes from *Muscadinia rotundifolia* and American vine species that showed resistance to downy and powdery mildew. He began back-crossing *Muscadinia* with vinifera varieties, beginning in 1974. In 2000 INRA-Colmar took over this important work and it became the ResDur program.

Two varieties that came from this initiative, Regent and Bronner, were particularly promising.

These had resistance derived from both American and Asian varieties, and the researchers began trying to "pyramid" resistance genes. This involves having multiple copies of the genes conferring resistance to downy and powdery, which makes it much more difficult for the microbes to develop their own resistance.

Bronner is a cross between Merzling and Geisenheim 6494. The crossing was made in 1975 and the variety released in 1997. It has two genes conferring powdery resistance (Ren3, Ren 9) and two conferring downy resistance (Rpv10, Rpv33). Merzling is itself a cross between Freiberg 379-52 (which is a cross of Riesling and Ruländer), and Seyval Blanc.

In June 2016, four ResDur varieties were included in the French catalog of permitted varieties. These were Muscaris Blanc, Souvignier Gris, Monarch Noir, and Prior Noir. In addition, seven more varieties were temporarily included: Bronner, Johanniter, Solaris, Saphira, Cabertin, Pinotin, and Divico. Three further varieties were given approval, but only if they changed their names to take out the famous name appropriation, which the French do not allow: Cabernet Blanc, Cabernet Cortis, Cabernet Jura. The appeal of using one of the names of an existing popular variety is clear to see, and it is allowed in some countries but not others. This is an issue that these new varieties have to address.

Section 2
In the Winery

8 Yeasts and bacteria

FERMENTATION: HOW YEASTS AND BACTERIA MAKE WINE

Yeasts don't get enough credit. When it comes to wine, grapes get all the glory. But without yeasts, all we'd have is grape juice. The choice of yeast, or indeed the decision of whether to use indigenous yeasts or cultured strains to carry out fermentation, is an important part of winemaking. Not only do yeasts convert sugar to alcohol, they are also able to influence the flavor and aroma of the final wine. Of the estimated 1,000 or so volatile flavor compounds in wine, at least 400 are produced by yeast. Take some freshly crushed grape must. It doesn't smell like much of anything. In contrast, the fermented wine is often rich in aroma and flavor, thanks to the action of yeasts.

We have agreed that yeasts are undervalued, so this chapter seeks to redress the balance a little, looking at their role in winemaking and addressing the science of fermentation. I'll also cover the controversy that surrounds wild-yeast ferments and attempts to engineer beneficial traits into yeast strains by genetic modification. And if yeasts are the underdogs, bacteria are almost completely ignored. But they, too, are important in modifying the flavor of wine in those situations where malolactic fermentation takes place—and this includes practically all red wines and a good subset of whites. So, we'll be taking a look at bacteria, too.

MICROBES AND WINE: SOME CONCEPTS

Until the 19th century, fermentation must have seemed a mysterious process to winemakers. It wasn't until the now-famous studies of Pasteur in the latter half of that century that yeasts were shown to be directly responsible for converting the sugar in grape must into alcohol. He correctly surmised that the particular yeasts carrying out the fermentation could influence the flavor of the wine. As Émile Peynaud, the famed wine scientist from Bordeaux, author of influential works on winemaking and wine tasting pointed out, "Before his time, good wine was merely the result of a succession of lucky accidents."

You can't see yeasts. Like bacteria, these unicellular fungi are far too small to be seen with the naked eye. Humans are tremendously visually centered, and it is perhaps because of this that we've found it hard to understand and be comfortable with the microbial world that surrounds us. In medicine, this has no doubt contributed to the current antibiotic crisis. Because we are unable to appreciate that microbes are everywhere, the message that there are good bacteria as well as pathogenic ones has been hard to handle. We are much more comfortable with the idea that bugs are bad and we should zap them all. This sort of attitude has encouraged the irresponsible overuse of antibiotics, and led to the development of widespread antibiotic resistance, a major threat to human health.

So, before we get to the gritty details, it is probably worth taking a conceptual look at microbes in winemaking. Yeasts and bacteria are ever present in the winery. Even in a spotlessly clean environment there will always be some receptive surface, such as an uneven soldered joint in metal pipework, where microbes can hide. Barrels are particularly receptive to microbes because the structure of wood means that it is almost impossible to sterilize it completely. Because a potential source of inoculation is just about ubiquitous, yeasts and bacteria simply need the right sort of environment, and they will begin to grow.

Grape must represents a sugar- and nutrient-rich medium that is ideal for the growth of certain microbes, although with its strong osmotic potential it also presents challenges for these microbes to overcome. As the must ferments, it changes, and its suitability for one species or strain wanes as its suitability for another develops.

Let's illustrate this in picture terms. Take the side of a mountain. At the base vegetation is lush and plentiful. The environment here suits a wide variety of organisms. Move a short way up and the change in climate affected by the altitude difference will mean that a different population of plants will prevail. This will continue up the mountainside, until conditions are such that toward the summit plants can no longer establish themselves. It is something like this in fermenting wine. Create the right conditions and you can select for the population of organisms that you want to be growing at that particular time.

Winemakers tend to concentrate on eradicating rogue organisms from the winery. But they might be better off concentrating on ensuring that their musts and developing wines that represent ideal habitats for the sorts of microbes they want to encourage, while at the same time not neglecting winery hygiene. To quote Peynaud, "The winemaker should imagine the whole surface of the winery and equipment as being lined with yeasts."

Microbes have short generation times, so they can be fiercely competitive. If conditions suit one yeast or bacterium a little more any others, it will rapidly outpace the competition and establish itself as the primary fermenting organism. Winemakers have to make sure that the musts they are working with give a competitive advantage to the sorts of bugs they'd like to see growing.

"NATIVE" YEASTS

Yeasts are widespread not only in the winery, but also in the vineyard. They spend winter in the upper layers of the soil, spreading to the vines during the growing season via aerial transmission and insect transfer. They colonize grape skins during the maturation phase, although they never reach very high levels on intact grapes. Contrary to popular opinion, the bloom on the surface of grape skins is not made up of yeast populations, but rather a waxlike scaly material that doesn't harbor many fungi.

Only a limited number of yeast species are present on grapes, the so-called "native" yeast populations. These include: *Rhodotorula,* the apiculate yeasts *Kloeckera apiculata,* and its soporiferous form *Hanseniaspora uvarum* (the most common by far), and lesser amounts of *Metschnikowia pulcherrima*, *Candida famata*, *Candida stellata*, *Pichia membranefaciens*, *Pichia fermentans*, and *Hansenula anomola*. Also present may be potential spoilage organisms, such as *Brettanomyces*.

Yeast nomenclature can be a confusing business, with various synonyms in common use for the same bug. This isn't surprising, because until the development of molecular methods for typing yeast strains and species, it was very hard to tell all of them apart. Until recently, the prevailing opinion was that the main wine yeast, the alcohol-tolerant *Saccharomyces cervisiae*, was rare in nature. Attempts to culture *S. cerevisiae* from the skins of grapes have largely proved unsuccessful. The only way its presence could be demonstrated is by taking grape samples and placing them in sterile bags, crushing them under aseptic conditions and seeing what happens. But modern sequencing techniques that do not rely on culturing have also shown beyond doubt that *S. cerevisiae* is present in the vineyard and on grape skins. At mid-fermentation, *S. cerevisiae*, which is a minor population on grape skins, represents almost all the yeasts isolated. In a few cases no *S. cerevisiae* is present and apiculate yeasts do the fermentation.

CULTURED OR SPONTANEOUS FERMENTATIONS?

So, you have your grapes and want to make some wine. You want to start a fermentation. There are two approaches open to you. Traditionally, the only option would have been to crush the grapes and leave the yeasts already present to get on with it. This is known alternatively as a "spontaneous," "wild-yeast," "indigenous- yeast," or "native-yeast" fermentation. Since the 1960s, with the ready availability of cultured strains of *S. cerevisiae*, winemakers now have the choice of inoculating the must with a starter culture of their preferred yeast. Estimates are that worldwide, some 80 percent of wine is produced by inoculated fermentations, but this is a tricky figure to verify and is changing all the time.

The choice between native and cultured yeasts has opened up a philosophical divide between winemakers who want to bring fermentation under control and those who prefer to leave things to nature. These days there is a loose kind of Old World–New World divide, with the former largely preferring to use indigenous yeasts (at the high end, at least) and the latter relying on cultured strains, although this divide is far from an absolute. I remember when I was first drinking wine in earnest in the mid-1990s that Chilean winery Errazuriz (with Kiwi Brian Bicknell of Marlborough winery Mahi the then winemaker) was marketing their Chardonnay as "wild ferment" in the UK retailer Oddbins. It seems strange to think that back then wild ferment was deemed a rare, potentially risky practice. Now it is extremely common for high-end wines. For example, the use of wild fermentations has increased a lot over the last decade or so in Australia's Barossa Valley. "The evolution has been nothing short of outstanding," says Kevin Glastonbury, winemaker with Yalumba. "More people allow wild yeast to do the fermentations now than the use of cultured yeast. Twenty years ago we weren't a hundred percent cultured yeast, but we were

close, but now we are over sixty–seventy percent indigenous yeasts. Most of the winemakers I talk to do it with many of their better wines."

What happens during a spontaneous fermentation? Because the initial inoculum of yeasts from the winery environment and grape skins is quite low, things can take a while to get going. This introduces an element of risk: if bugs such as *Acetobacter* (the acetic acid bacterium that turns wine to vinegar) establish themselves before the fermentative-yeast species, then the wine will be at risk of spoilage. Also, there's no guarantee that the native yeasts that become established will do a good job. Like *S. cerevisiae*, all the various native yeasts exist in many different strains, some desirable, others not. With a spontaneous fermentation you take what you are given.

Apiculate ethanol-sensitive yeasts, such as various species of *Hanseniasporia* (and its anamorph [asexually reproductive] form *Kloeckera*), which is the dominant yeast found on grape berries, usually dominate the initial stages of a wild ferment. As these decline yeasts, various species of *Candida* take over (*Candida stellata, C. pulcherrima* [and its teleomorph, the sexually reproductive form *Metschniokowia pulcherrima*], and *C. colliculosa* [and its teleomorph *Torulaspora delbrueckii*]). It has been estimated that in an uninoculated ferment, as many as 20–30 strains participate. But as alcohol levels reach 4–6 percent most native species can't take it, and the alcohol-tolerant *S. cerevisiae* will take things onward from here. The only two non-*Saccharomyces* yeasts that can tolerate higher alcohol levels are *C. stellata* and *C. colliculosa*. Other species of yeasts may be present in wild ferments. These include *Cryptococcus, Debaromyces, Issatchenkia, Lachancea, Pichia, Rhodotorula,* and *Zygosaccharomyces*.

There are twists to this story, however. Most winemakers will add some sulfur dioxide on crushing. This will slant things in favor of *S. cerevisiae* and the more robust of the native species, eliminating some of the more problematic wild yeasts and spoilage bacteria, which tend to be more sensitive to the microbicidal actions of sulfur dioxide. Temperature also affects the balance of yeast species in the fermentation. Cooler temperatures (below 14°C/57°F) favor wild yeasts, such as *Kloeckera*, whereas higher temperatures shift things in favor of *S. cerevisiae*. Added to this, as harvest gets underway, the winery equipment will be a ready source of inoculum, and fermentations will get going a lot faster, with *S. cerevisiae* establishing itself sooner. Studies have shown that after a few days of harvest operations, half of yeasts isolated from the first pumping over of a spontaneously fermented red-grape tank are *S. cerevisiae*. Aside from the actual properties of the wild yeasts themselves, spontaneous ferments cause a delay the onset of vigorous fermentation. This will allow oxygen to react with anthocyanins and other phenolics present in the must, enhancing color stability and accelerating phenolic polymerization.

Why take the risk of a spontaneous wine fermentation? In many cases the motivation will be ideological—this is the traditional approach in certain areas and there is a reluctance to adopt alternative methods, or a disbelief in the integrity or efficacy of these alternative methods. Others do it for quality reasons because native yeasts are thought to produce wines with a fuller, rounder palate structure, and the ferments tend to be slower and cooler, burning off fewer aromatics. There is also a cost saving because cultured yeast has to be paid for.

It should be pointed out that even where cultured strains of yeast are used, wine must is not a sterile medium, and even though a good dose of sulfur dioxide will kill off the more susceptible bacteria and yeasts already present, some indigenous strains will likely play a small role in the fermentation.

Some winemakers establish a *pied de cuve*, which is a kind of starter fermentation. Before the main harvest, they will pick some grapes, crush them, and let fermentation begin. This is acts as a type of insurance policy against anything bad happening. Not only does it get fermentation started more rapidly, when a few *pied de cuve*s are established, they give the winemaker a chance to select a ferment to inoculate with that smells nice.

AN IDEOLOGICAL DIVIDE: WILD YEASTS VERSUS CULTURED YEASTS

Nicolas Joly sums up well the objection that some winemakers have to using cultured yeasts to initiate fermentation. "Re-yeasting is absurd. Natural yeast is marked by all the subtleties of the year. If you have been dumb enough to kill your yeast you have lost something from that year."

He's not alone. Many winemakers, particularly those in the classic Old World European wine regions, see the use of cultured yeasts as unnecessary and even plain wrong. They argue that the native yeasts present in the vineyard are part of the terroir. "We are very big fans of wild-yeast ferments," says Pierre Perrin, of Château de Beaucastel in France's Châteauneuf du Pape region. "It can be risky if the fermentation doesn't begin quickly. But if you do a *pied de cuve* two or three days before your first harvest day, you can mix this with your crop after a few days of maceration and then the fermentation will go. The key is to have a fermentation departure that isn't too fast (which eliminates the maceration process) or too slow (which risks acetic problems)." However, most New World winemakers, and a growing band of Old World producers, now initiate winemaking with cultured yeasts, seeing the control of fermentation parameters this affords as being key to quality.

The choice of specific yeast strain is also seen as an important winemaking decision, because the various properties of the yeast can be chosen to complement or add to the wine style being made. Yeast expert Sakkie Pretorious points out that the outcome of spontaneous fermentation is highly unpredictable and describes the risk involved as "potentially staggering".

MATT GODDARD'S WORK ON WILD YEASTS

New Zealand-based yeast researcher Matthew Goddard has shown that in a natural ferment, *S. cerevisiae* engineers its environment to give itself a competitive advantage through the production of alcohol and heat. Fermentation of sugars is actually less energetically favorable than respiration of these sugars (which requires oxygen), but even in the presence of oxygen *S. cerevisiae* will still choose to ferment these sugars, producing alcohol in the process. Goddard speculates that because *S. cerevisiae* is a specialist at consuming ripe fruits, it chooses to make ethanol to protect this valuable food resource. The ethanol acts as an antimicrobial, reducing competition, and also deters most vertebrates. Thus *S. cerevisiae* is choosing a less efficient pathway, which you'd have expected evolution to eliminate, unless there was some collateral benefit. This production of ethanol is therefore an example of niche construction.

Goddard's work makes use of new genetic tools that are able to differentiate between different strains of yeasts. "It is not an easy thing to do. We use exactly the same technique that is used in forensic criminology situations—genetic fingerprinting—to distinguish different strains of *S. cerevisiae*. This affords us a powerful insight into the variation and relatedness of different strains that we have isolated from different areas."

His most important findings relate to a series of studies that took place in West Auckland, in New Zealand. New Zealand is geographically remote and was only inhabited by humans as recently as 700 years ago. The human introduction of wine occurred just 100 years ago.

In the first study, Goddard looked at a spontaneous Chardonnay fermentation at boutique winery Kumeu River, when 800 random isolates were taken from the ferment. Initially, *S. cerevisiae* was present at very low frequency (1/1,500 of the population), but by day 11 it dominated the fermentation. Using genetic fingerprinting, Goddard found 88 different genotypes from 380 isolates of *S. cerevisiae*. Using some statistical techniques to analyze the genetic data, he concluded that there were around 150 different genotypes present in the fermentation, and these derived from six very distinct subpopulations. But were these "wild" populations of *S. cerevisiae* actually escaped commercial strains?

Goddard and his team compared them with a large database of commercial yeast strains and found that they were quite different. They then combed the winery, sampling winery equipment and walls before harvest, and failed to find any *S. cerevisiae*. The conclusion was that the strains in the wild ferments they analyzed were brought into the winery with the grapes. "For a long time the impression in the industry has been that *S. cerevisiae* doesn't live in the natural environment; it only inhabits wineries, and that these are escaped domestic/commercial strains," says Goddard. "When I was at the Oregon Wine Board in February to talk about some of this, many of these people had been schooled through the UC Davis system and this is what they had been taught. They were astounded by our results."

In the next phase of the experiment, Goddard and his coworkers sampled soil, bark, and flowers from Matua Valley vineyard, which is just 6km (about 4 miles) from Kumeu River and which is surrounded by bush. They found 122 different colonies of *S. cerevisiae*, with 22 different genotypes. Two were found in vine bark, two in buttercup flowers, and the remainder in the soil.

None of the Matua Valley genotypes matched any commercial isolates, nor did they match the isolates in the Kumeu River ferment. The inference is that it is these local strains, from soil bark and flowers, which end up carrying out the indigenous fermentations. It makes sense that *S. cerevisiae* must be living somewhere in the environment, because ripe fruit is only available as a habitat for a couple of months a year at most. But how are these strains spreading among the soil, bark, and flower niches, and then getting onto the fruit? The obvious explanation seems to be by insect transfer, so Goddard took 19 samples, over a five-month period, from an apiary with hives near both West Auckland vineyards that he has studied. From the 67 colonies of bee-borne *S. cerevisiae* analyzed, two were almost identical to isolates from the Matua Valley vineyard.

But one question remained. How similar were these New Zealand isolates to yeasts found in other countries? The answer was they were pretty unique. On average, the New Zealand isolates were found to share less that 0.4 percent of their ancestry with international strains of *S. cerevisiae*, and therefore were singular populations. "This was the first demonstration that strains in spontaneous ferments derived from the local environment," says Goddard, "and these strains appear to be unique (or at least distinctive) when compared with strains from elsewhere on the planet."

There was a further twist to the story to come, however. Kumeu River winemaker Michael Brajkovich suggested that Goddard should examine a new French oak barrel, to see whether there was any possibility that yeasts could be transferred internationally in this way. So, they looked at a new barrel imported from Chagny in Burgundy and found 50 different isolates of *S. cerevisiae* with 40 different genotypes. This is the first time that *S. cerevisiae* has been shown to be present in new oak barrels. Significantly, one of these isolates was shown to be the same as one of the genotypes present in the Kumeu River wild ferment. "This was a bit of a shock," says Goddard. "We were pretty stunned to see this. Again, it is applying ecological population biology techniques to this system in a rigorous way and bringing the right tools to the question."

Goddard has extended his study to other wine regions. One of these is Central Otago, where he has looked at yeast populations at Felton Road. Nigel Greening of Felton Road reports that the results show exactly the same thing as found at Kumeu River. The yeasts carrying out fermentations are local. "The biggest single cohort is unique to place, and this is 30–35 percent of the yeasts," says Greening. "The second-largest cohort tracks back through the barrels to the forests the oak was grown in. This is really interesting for Chardonnay if you are barrel fermenting. Then for the cohorts after this it gets harder to define where they came from. Just as with Kumeu River, Matt Goddard couldn't find a single strain that tracks back to a laboratory yeast, so we were clean."

One interesting question is whether systemic fungicides have any effect on the microbial populations on grapes, and this is a topic that Goddard is currently looking at with a Master's student. "We know a bit about what Mancozeb and copper oxychloride do to 'nasty' fungi, but what do they do to good fungi? There is a huge amount we could investigate. The anecdotal evidence from winemakers is that if you spray, it is harder to conduct a spontaneous ferment. Killing off the microbes in the vineyard, be they yeast or otherwise, has a knock-on effect."

Goddard's studies are hugely interesting and have important implications for how we see microbial populations as part of terroir. They also answer some long-disputed questions in wine microbiology. "If different regions harbor different populations of wine yeasts, then the use of these region-specific yeasts in winemaking means that the resulting wine more faithfully reflects the sense of place—or terroir—of the wine," says Goddard.

OTHER WILD YEAST STUDIES AROUND THE WORLD

Studies from elsewhere, however, have suggested that the actual picture is more complex. There have been a series of research investigations in the Okanagan Valley, in Canada's British Columbia. In the first, Master of Science student Jessica Lange looked at three different wineries, and compared spontaneous and inoculated fermentations in each of these regions, looking at four different stages of fermentation. In the younger winery (12 years old at that time) there was a wider diversity of species taking part in all the fermentations, but in the two older wines where there was a longer history of inoculating, most of the fermentations were carried out by commercial strains that had become resident in the winery. Even where the fermentations were inoculated with a commercial yeast, there were some instances where the resident strains displaced the inoculated strain and carried out the fermentation. However, all of the ferments—inoculated and wild—were treated with a 40mg/l dose of sulfur dioxide before they got going. This is usually done to sway things in favor of *S. cerevisiae*, rather than wild yeasts and bacteria, which have lower tolerance to sulfur dioxide, and while it's not a massive dose, it might be making life a little difficult for the wild yeasts.

In another study, researchers looked at Pinot Noir fermentations over two vineyards at Stone Boat winery in the Okanagan, from three different vineyard blocks. They identified 254 different strains of yeast, including 100 that were commercial strains. This was quite a high proportion of commercial strains, but there were enough noncommercial ones for the authors to conclude that the Okanagan probably has its own indigenous yeast strains.

Elsewhere the picture seems very mixed. Where people have looked, winery resident strains do seem to be quite invasive. Researchers have found that the surfaces in wineries harbor lots of yeasts, even when the winery has been cleaned quite carefully. And when fermentations are taking place, the air in the winery is full of yeasts. As a result, many studies report spontaneous fermentations being taken over by winery-resident commercial strains. But others are more in agreement with the results from Matt Goddard in New Zealand. One Austrian study looked across four regions and found that a wide diversity of strains of *S. cerevisiae* carried out wild ferments,

and they were all unique to Austrian vineyards.

Then again, a three-year study from two regions in Italy seemed to indicate that the effect of vintage on yeast populations overcame the terroir effect. Across 14 sites in Franciacorta and Oltrepò Pavese, of the 270 isolates of *Saccharomyces*, they identified 47 as commercial strains. They found a high diversity of yeasts, but the vintage conditions seemed to affect the diversity more than the geography, arguing against yeast populations being a fixed aspect of terroir. This is partly supported by a study looking at bacteria and fungi present on grapes in California, where the authors found an effect of climate. But they also found a varietal pattern (different grape varieties have different populations), and a link between geography and the different microbial populations.

CULTURED WILD YEASTS

What about winemakers who want some control over their fermentations, but like the qualities provided by cultured yeasts? In recent years, a number of cultured non-*Saccharomyces* yeasts have become available.

The yeast company Lallemand has a strong research program on the aromatic potential of non-*Saccharomyces* yeasts, combined with the optimization of the production of these yeasts in dry form. This allows winemakers to take advantage of fermenting with non-*Saccharomyces* wine yeasts while maintaining control over the fermentation process. The yeast *Torulaspora delbrueckii* strain 291 is now available to the winemaker in combination with specific *Saccharomyces* yeast in sequential inoculation, which results in wines with a distinct sensory profile and mouthfeel.

Other yeasts are also being investigated, such as *Hanseniospora orientalis, Schizosaccharomyces pombe, Metschnikowia pulcherrima, Hansenula anomala, Candida stellata,* and *Pichia anomala,* and also *Lachancea* (formerly *Kluyveromyces*) *wickerhamii* to control the development of

Brettanomyces yeast through their production of specific compounds. All those non-*Saccharomyces* yeast are found in the natural flora of wineries and vineyards, similarly to cultured *S. cerevisiae* used for alcoholic fermentation.

Chr. Hansen, another yeast company, has a reasonably broad portfolio of non-*Saccharomyces* products. "Prelude" is a single species culture of *Torulaspora delbrueckii*, and they also offer blends. "Melody" and "Harmony" are both blends of *Lachancea thermotolerans*, *Torulaspora delbrueckii*, and a strain of *S. cerevisiae*. "Rhythm" and "Symphony" are both blends of *Lachancea* and *S. cerevisiae*.

What effect do these cultured non-*Saccharomyces* have? One study suggests that *Torulaspora delbrueckii* enhances the complexity and fruity notes of wines. It is also a low acetic acid producer and so is often used in sweet wines.

Lallemand have a strain marketed as Biodiva, which is a high producer of polyols, which are sugar alcohols. Best known of this group is glycerol, but it also includes the C5 and C6 polyols arabitol, ribitol, sorbitol, mannitol, and xylitol. When their levels are elevated they change the mouthfeel of the wine giving a perception of sweetness. *Hanseniaspora vineae* gives fruity and flowery aromas.

Lachancea thermotolerans is of great interest because it is able to reduce alcohol slightly and increase acidity, by turning sugar into lactic acid. Lallemand have commercialized a strain selected from Rioja as Laktia, and according to their data, it can produce 2–9g/l of lactic acid. It significantly increases total acidity (TA) and decreases the pH, and *S. cerevisiae* is inoculated 24–72 hours later depending on how much lactic acid is wanted. One issue is that significant lactic acid production can make malolactic fermentation tricky, so there needs to be careful selection of the bacterial strain. Chr. Hansen have a strain of this, which they market as Concerto.

Metschnikowia pulcherrima is thought to reduce alcohol levels and enhance varietal aromas. In a paper from 2015, Contreras and colleagues showed that in musts inoculated first with *M. pulcherrima* and then finished with a strain of *S. cerevisiae*, alcohol levels were lower, by as much as 1.6 percent. *M. pulcherrima* is also used before fermentation as a bioprotective agent, because it is able to consume oxygen from the must. Lallemend sell a strain called Initia, which is used to protect the must of white and rosé wines from oxidation pre-fermentation.

Some important flavor compounds produced by yeasts during fermentation

Class of compound	Examples	Notes
Acetate esters	Ethyl acetate	Esters are largely responsible for fruity characteristics of wines. They hydrolyze during aging, but often remain at concentrations above threshold. Synergy between the esters present in wine determines how they are perceived
	2-methylpropylacetate	(also known as isobutyl acetate)
	2-methylbutylacetate	(also known as amyl acetate)
	3-methylbutylacetate	(also known as isoamyl acetate)
	Hexylacetate	
	2-phenylethylacetate	
Branched chain esters	Ethyl-2-methylpropanate	These are among the most powerful esters
	Ethyl-2-methylbutanoate	
	Ethyl-3-methylbutanoate	This has an aroma like strawberries, which contributes to the fruity character of some red wines
Higher alcohols (also known as fusel alcohols)	2-methylpropanol	(also known as isobutanol)
	2-methylbutanol	(amyl alcohol)
	3-methylbutanol	(isoamyl alcohol)
	2-phenylethanol	This has a pleasant roselike aroma
Volatile sulfur compounds	Hydrogen sulfide (H$_2$S)	The most common volatile sulfur compound in wines—smells like rotten eggs
	Ethanethiol (ethyl mercaptan)	Aroma of rotten onions or burnt rubber at threshold levels; at higher levels it is unpleasantly fecal
	2-mercaptoethanol	Aroma like a barnyard
	Methanethiol (methyl mercaptan)	Rotten cabbage

BIOPROTECTION

In July 2020, I visited Château Hospitalet, which is one of the châteaux owned by Gérard Bertrand. There are two sides to this impressive wine empire. On the one hand, they make high-end wines from their 16 châteaux, which are farmed biodynamically. On the other, they make large quantities of more affordable wines together with partner wineries. Included in this is the three-million annual production of a range called Naturae, which are made with no added sulfites.

I tasted through the Naturae wines all the way back to 2011 with winemaker Stéphane Quérault, who is in charge of all the Bertrand partner winery projects. They were impressive and had aged well. How do they do this? One of the techniques used is bioprotection; another oenologist, Stéphane Yerle explained the concept (he is a consultant who works with wineries such as Bertrand, Maison Ventenac, and Famille Fabre). The idea is making use of competition and using up any nutrients that otherwise could support growth of unwanted microbes. A typical protocol would be to inoculate the must with bacteria (for example *Lactobacillus plantarum* might be added since this does not produce any volatile acidity [VA]), then *Lachnacea thermotolerans,* which does some fermentation and raises the acidity. Then a strain of *S. cerevisiae* will take over and as soon as alcoholic fermentation is complete another strain of bacteria, usually *Oenococcus oenii*, is added to finish off the remaining malolactic fermentation. The idea is that this creates competitive exclusion: strains of microbes are chosen that can outcompete undesirable microbes and produce compounds that inhibit any potentially problematic yeasts and bacteria. Some selected bacteria have a direct inhibitory effect on *Brettanomyces* growth, for example. There is more on this in chapter 13, see page 165.

SELECTING FOR YEAST STRAINS: DESIRABLE PROPERTIES

Yeasts do a whole lot more than just convert sugar into alcohol. They are responsible for the metabolic generation of many wine flavor compounds from precursors in the grape must. Because of this, the use of strains of cultured yeast with specific properties has become an important winemaking tool, although one that is not universally welcomed—some traditional producers see this as a way of cheating.

Wine microbiologists see the development of yeast strains with enhanced abilities as an important goal in furthering wine quality. There

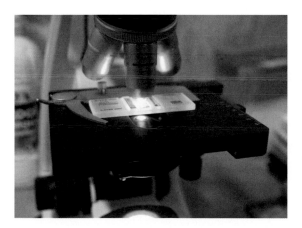
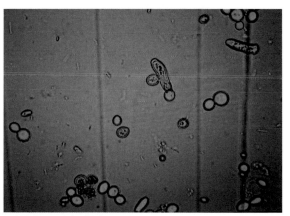

Above Yeasts under a microscope. This is at Lapierre, a natural-wine producer in France's Beaujolais wine region. They look at the populations growing in their wines after alcoholic fermentation has finished to check that nothing bad is present.

are two ways of doing this. The first is by more "traditional" genetic techniques that don't involve the direct introduction of new genes. These include selection of variants (choosing the best of a range of natural genetic variants); mutation and selection (using a mutagen to increase the frequency of genetic variation and then selecting for those mutants with enhanced properties); hybridization (mixing together different species); and spheroplast fusion (a special way of joining together yeast cells to produce progeny with enhanced properties). The second is by transformation: the precise introduction of new, specified genes into the genome of the yeast strain of interest. This is also known as genetic modification (GM).

Both have their benefits and drawbacks. If the trait of interest is polygenic (multiple genes are involved), then non-GM methods are the best way to select for this. However, they are less precise, which means you run the risk of losing the beneficial traits of the starting yeast strains. GM methods are much more precise and elegant, but things get complicated if more than one gene is involved, and there is still the huge hurdle of public antipathy. It is likely that both types of strategies will prove important.

What are some of the desirable properties that microbiologists would like to engineer into yeasts? There's actually a fairly long list, summarized on the page opposite.

It is clear that there is a lot of scope for the manipulation of yeast strains to enhance wine quality. However, such developments are unlikely to appeal to the traditional wine producers who see the wild yeasts that dominate the early phases of spontaneous fermentations as part of their terroir, and an important factor in fine-wine production.

FLOR

What is flor? It is something that grows on the surface of wine: a film of yeasts. And it alters the flavor of the wine in interesting ways. Currently in fashion, it is a subject that is worth finding

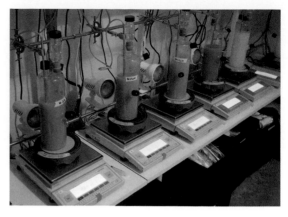

Above Experimental fermentations testing the performance of different yeast strains. The scales allow fermentation progress to be tracked because as carbon dioxide is lost the weight goes down.

out about. We can only speculate about how the effects of flor were first discovered. Perhaps somebody left a barrel incompletely filled. The air in the headspace at the top of that barrel contains oxygen, which would have interacted with the wine. And before long, a thin film of what looked like mold would have grown on its surface. This would have been rather an alarming development. Perhaps the mold might have been growing quite a while before it was detected, and the winemaker— scared that the wine might have gone bad—may then have tasted the wine with trepidation. It must have been a great relief for them to find a wine that instead of being ruined was fresh and had some really interesting characteristics: a salty tangy bite, lovely savoriness, and no signs of oxidation at all.

Two wine regions have become famous for flor. The first is Jerez in Spain's Andalucia region, where the biologically aged sherries Fino and Manzanilla develop under flor. A third style, Amontillado starts its life under flor then develops oxidatively afterwards. The second is France's Jura region, where Vin Jaune also develops this way. There are other places with a tradition with flor-aged wines, but they are less well known. There now seems to be much wider interest in making wines this

Goals for genetic improvement of wine yeast strains

Fermentation performance	Fermentations are often faster than optimum, so lower temperatures often control them. But sometimes, though, they are sluggish, even becoming "stuck" with disastrous consequences for quality. Improved yeast strains could help here. Targets include greater stress resistance, improved grape-sugar and nitrogen uptake, resistance to high alcohol levels, and reduced foam formation. It would also be beneficial to have yeasts that can utilize the nitrogen sources in wine that they currently don't (this would help avoid stuck fermentations), and for them to be resistant to toxins produced by wild yeast strains.
Biological control of spoilage bugs	Spoilage microbes are a constant threat and are countered by the addition of sulfur dioxide. A new development has been the use of antimicrobial peptides and enzymes as an alternative to chemical preservation. It would be ideal if these could be synthesized directly by yeasts.
Processing efficiency	The fining and clarification of wine consumes time and resources and risks the removal of flavor components. Wouldn't it be great if yeasts could do this job? They could be engineered to secrete proteolytic and polysaccharolytic enzymes that would remove proteins and polysaccharides, which can form haze and clog filters. Another avenue of research is the regulated expression of flocculation genes, which would enable winemakers to encourage the yeast to enter into suspension for fermentation, and then settle quickly as a residue on completion.
Flavor and sensory qualities	Wine is a complex mixture of hundreds of different flavor compounds, many of which are synthesized by yeasts. Yeast-derived compounds can be both positive and negative in terms of flavor impact. Therefore, yeasts with a positive impact should be selected. Yeasts that possess enzymes that liberate color and aroma have been selected, as have those producing ester-modifying enzymes. Elevated alcohol levels are becoming an increasing problem in many regions and attempts have been made to produce yeasts that ferment to dryness at lower alcohol levels by diverting more sugar to glycerol production. Yeast strains are also being developed to adjust acid levels biologically.
Healthy properties	Sometimes wine contains elevated levels of undesirable compounds such as ethyl carbamate and biogenic amines. It is ideal to minimize the levels of these. Yeasts could also be developed with enhanced production of supposedly health-enhancing compounds, such as resveratrol.

way, even where it was never a local tradition. This is part of the general opening of minds in the world of wine that has also seen the return of clay (amphorae, *tinajas*, *talhas*, and *qvevri*) and a burgeoning interest in making wines naturally.

I remember well my first visit to Jerez. To experience a major wine region for the first time is always memorable. Also, what I had learned in books—that Palomino is an essentially neutral grape variety and that the vineyards here are all the same—was shaken up by what I saw on that visit. A return trip a few years later further convinced me that the future for this region involves a return to the vineyards and turning back the clock 50 years or more to before the expansion of vineyard area and the creation of "big Sherry." A careful reading of the history books shows that in the past, the best Finos and Manzanillas were not fortified at all. And now there are a small band of producers who are making unfortified biologically aged wines that are quite stunning. Of all the wines in the region, it is those biological Sherries, aged under flor, that are, to me, the most compelling wines.

All Sherry starts out as a dry white wine, known as *mosto*. In the old days, when all Sherry was fermented in barrels, the classification of wines would have been an immense task. These days, most is fermented in much larger stainless-steel tanks, and so it makes life for the cellarmaster a lot easier. Generally speaking, the lighter, more delicate wines from the best *pagos* (vineyards) are chosen for biological aging, and they are put into barrels that are deliberately only 80 percent filled, in order for the layer of flor to grow at the interface between the air and the wine. Most of these wines are fortified to around 15.4% alcohol, which is close to the limit for flor growth, but any lower and less desirable microbes might be encouraged to grow with the attendant risk of spoilage. Wines for oxidative aging are fortified to 18% alcohol, so nothing can grow in them. But there are some producers who are going back to the old method of drying the Palomino grapes for a day or so on small mats, called *asoleo*, to raise the sugar levels so that no fortification is needed.

Technically, the flor (or "velum," as it is also known) is a biofilm made up of the yeast

Above Flor growing on the surface of a Zibbibo wine at Brash Higgins, McLaren Vale, Australia. This is a strain of *Saccharomyces cerevisiae* that forms a velum and grows on the surface of the wine, feeding off glycerol, acetic acid, and alcohol and creating interesting flavors, as well as keeping the wine fresh.

Saccharomyces cerevisiae. This is the same species of yeast that carries out alcoholic fermentation, but it is a specially adapted version that is able to cope with the rather hostile conditions found in a Sherry cask. The challenges facing the yeast are quite extreme. The wine in cask is quite acidic, has high alcohol, very little sugar, and there is also usually around 30mg/l of sulfur dioxide present. But the flor yeast has adapted its cell wall so that it floats and is able to form a surface film where it is able to access the oxygen needed to metabolize the alternative carbon sources of alcohol and glycerol. It also has efficient antioxidant defenses to cope with the increased oxidative stress this lifestyle brings.

HOW DOES THE FLOR AFFECT WINE FLAVOR?

If the layer is thick enough, it protects the developing wine from the air present in the top of the barrel, which means that unlike an Oloroso, which turns brown, the Sherry under this layer stays fresh and pale in color. As the yeast feeds off alcohol, it produces acetaldehyde, which has flavors of nuts and apples and in this process the alcohol level of the Sherry is reduced by around 0.2–0.3% a year. Acetaldehyde levels in a Fino or Manzanilla are typically around 300–400mg/l, but can be as high as 800. Glycerol, another important wine component, is also consumed by the yeasts, and dips from a starting level of around 7g/l down to around 0.3g/l, which changes the mouthfeel, making the wine feel less viscous and fresher and lighter. There is also an increase in the levels of a compound called sotolon, which adds spicy, curry, nut notes.

Two factors help increase the thickness of the flor layer. The first is the way that Sherry is made in the solera system. New wine is being introduced into each barrel at various stages, using a special "canoe" device that doesn't disrupt the flor, and this brings with it fresh nutrients to keep the yeast cells growing healthily. The second is the cellar environment: cooler, more humid conditions make for a thicker flor layer. Consequently, the coastal towns of Sanlúcar de Barrameda and El Puerto de Santa Maria tend to have higher levels of flor, and there are also seasonal changes with thicker flor in spring and fall. Should the flor struggle or die, then the Sherry can be fortified further to 17 or 18% alcohol, and then it begins its second, oxidative journey. This is then an Amontillado.

How does flor affect flavor? I find it adds a distinctive tangy quality. There's some saltiness, too, and a lovely savory edge. The acetaldehyde tends to add some appley character, which you'd normally associate with oxidation, but in the context of a fresh, zippy white seems to work. Flor character is a bit like oak, in that it can be overdone. Some is good, but that doesn't mean more is better. Microbes make wine, but usually after fermentation (and malolactic fermentation) their job is over. The wonderful thing about flor-aged wines is that the microbes are involved all the way.

Whether or not flor appears is still a bit mysterious. If you are working in a region where it isn't traditional, often the first flor-aged wines are made as a sort of accident: a flor layer appears on the wine and then the winemaker decides to take a risk and let it carry on growing. It's becoming a legitimate style choice. Few have the courage to leave the wine under flor for as long as happens in Jerez or the Jura, but I've still had some impressive examples. In the future, it will be exciting to see more unfortified flor-aged wines coming from Jerez. The rules have recently changed so that it's now legal to call unfortified wines from the region Sherry, in a return to the past.

MALOLACTIC FERMENTATION

The term "malolactic fermentation" (MLF) is actually a bit of a misnomer, because what we're referring to is the transformation of one acid, malic acid, into another, lactic acid, with the release of carbon dioxide. This is just one of the effects that the malolactic bacteria have on wine, but it is the most important one.

The bacteria we are concerned with, known as lactic acid bacteria (often abbreviated to LAB), are present in the vineyard on the grapes, much like the yeast species that carry out spontaneous fermentations. The microbial population on the grapes gets slimmed down a little at crushing, because of the hostile conditions of grape juice, with high acidity and high sugar content. Yeasts are better at dealing with these conditions than bacteria, and the bacterial populations tend to tumble somewhat, especially when fermentation gets going.

The first mention of MLF is thought to be in a book from 1837 by Freiherr von Babo. In this he described a second fermentation occurring in some wines in spring, as the temperatures began to rise. This resulted in the release of carbon dioxide and renewed turbidity in the wine. Then, in 1866, the celebrated scientist Louis Pasteur isolated bacteria from wine for the first time. Pasteur considered, however, all bacteria in wine to be spoilage organisms.

A breakthrough came in 1891 when Hermann Müller postulated that acid reduction could be because of bacterial activity. This was a bold suggestion, because at the time this change in acid levels was thought to be because of tartaric acid precipitation. He went on to do some important work over the next couple of decades in collaboration with a fellow Swiss researcher called Osterwalder. In 1939, famous French wine scientist Émile Peynaud wrote an important paper on the role of malic acid in the musts and wines in Bordeaux, pointing out that the absence of MLF was a quality-limiting factor in these wines. "Not only is the acid make-up of the wine completely changed," he stated, "but [MLF] has an impact on the perfume of these wines and even diminishes the intensity of the color and changes its shade. It is not exaggerating to say that without malolactic fermentation, there would hardly be any great reds of Bordeaux."

The first time LAB were cultured was in the late 1950s. In 1956, Brad Webb was appointed winemaker at a new winery, Hanzell, which was founded by the wealthy US ambassador to Italy, J.D. Zellerbach. Zellerbach wanted to make Chardonnay and Pinot Noir that rivaled the classic wines of Burgundy. The new winery was well suited for making high-quality Burgundian-style wines, and Webb had at his disposal an array of shiny new tanks, casks, and barrels. But there was a problem. The first Pinot Noirs that Webb made in the new facility would not do MLF. At this time, a growing awareness was developing of the importance of MLF for high-quality red wines. Sometimes it started of its own accord; sometimes it didn't. Sometimes it finished quickly and easily; other times it dragged on, only to start again at a later stage.

While yeast cultures were beginning to become available, no one had yet successfully cultured the LAB responsible for the conversion of malic to lactic acid. So, Webb had to resort to several techniques to try to get this fermentation to start. Could the reluctance of his ferments to do MLF be because of the lack of suitable bacterial inoculum in the winery? He tried introducing wine that was undergoing MLF into his tanks, but it didn't work. He turned to a scientist at UC Davis who had an interest in bacteria, John Ingraham. Webb offered his winery as an experimental setting where Ingraham could do trials: after all, in almost all other wineries MLF took place. So, it should be of interest to study a winery where such a fermentation had never occurred. Ingraham was intrigued and was willing to use Hanzell as a negative control.

Research on LAB, which are also important in the production of other foodstuffs, had begun to flourish ever since the serendipitous discovery by a dental researcher in Chicago that adding tomato juice to the Rogosa medium commonly used for making microbial cultures allowed these previously tricky bacteria to grow more easily in the laboratory. Tomato juice contains panthenoic acid, which is an important growth factor for these bugs. Ingraham had isolated 50 strains of LAB from samples of dry wines and lees, from several Californian wineries. Of these, with the help of Webb, he selected one that showed the most promising characteristics, and this was given the name ML34.

Although this was kept secret at the time, it came from a large redwood tank of Barbera in the Louis Martini winery in Napa. In 1959, the two took ML34 to Hanzell to study it in winery conditions. After some trials on a small scale, to their great excitement they got this LAB to carry out a MLF in winery conditions. To their knowledge, they were the first to do this. But as so often happens in science, a couple of other groups had been working on the same problem elsewhere. In France, Émile Peynaud had achieved this with his colleague Simone Domercq, and a short time before, a Portuguese group had also been successful. But this doesn't take anything away from the achievements of Webb and Ingraham, who published the results of their work in 1960.

It is quite common for winemakers to add some sulfur dioxide at crushing. This is toxic to both yeast and bacteria, but bacteria are more susceptible, and so this addition tends to clear the field for the yeasts to do their thing. Typically, the population of LAB present in must is 100–10,000 viable cells per milliliter, expressed as CFU/ml (where CFU stands for colony-forming units, referring to cells that can form a colony when plated onto a culture medium). This initial population dips as fermentation progresses because of the increasingly hostile conditions encountered—alcohol levels rise and nutrients are depleted from the must. Also, some yeast strains are believed to release antimicrobial substances to eliminate the microbial competition. By the end of alcoholic fermentation, there will usually be only 10–100 CFU/ml of these LAB remaining, but they are ready and waiting to start a rearguard action.

By this stage other things will have shifted in favor of bacterial growth. The nutrients, depleted during the active phase of fermentation, will have restored their levels a degree, and free sulfur-dioxide levels will have dipped to around zero. The other factor important for bacterial growth is temperature, and if this increases a little, for example, through the cellar warming up in spring, then bacterial populations can build up again.

There are four genera of LAB important in wine. The main one is *Oenococcus* (a single species, *O. oeni*). There are also 12 species of *Lactobacillus*, three species of *Pediococcus*, and a single species of *Leuconostoc*. As with yeasts, it is not just the species that matters, but also the strain: *O. oeni*, for example, shows important strain differences. Many malolactic fermentations are spontaneous, but since the 1980s it has been possible to buy cultured strains of *O. oeni* to inoculate with and start MLF to order.

INOCULATING WITH LACTIC ACID BACTERIA

Traditionally, MLF has been rather unpredictable. It happens when it happens, and sometimes it happens after bottling, with disastrous consequences. To initiate the process, the population of LAB has to reach a threshold level of one million viable cells per milliliter (10^6 CFU/ml). Factors influencing this population growth include pH (it happens more easily at higher pH, but then the species and strains taking part will likely be a little different), the nutrient status of the wine, the temperature, and the strains of bacteria present. In many classic European regions, harvesting takes place in fall and cellars become increasingly cold as fermentation finishes, such that there is a long delay before malolactic fermentation, which takes place the following spring as the cellars warm up. This means a delay of several months between the completion of alcoholic fermentation and the onset of the malolactic conversion.

This delay between the end of alcoholic fermentation and the onset of MLF can be a risky time for red wines. This is because the wine is relatively unprotected from microbial growth because free sulfur dioxide levels are low or zero, creating ideal conditions for the growth of the spoilage yeast *Brettanomyces*. Winemakers are not able to use much sulfur dioxide at this stage because they risk inhibiting MLF altogether. This is why some choose to inoculate with bacterial cultures. Once MLF is finished, it is possible to add sulfur dioxide to protect the wine. But it's a complicated business, because wines respond differently to the ingress of oxygen that barrel maturation permits depending on whether or not free sulfur dioxide is present at appreciable levels. *Élevage* is a complicated business, and perhaps as much an art as a science.

Another factor is whether or not to complete MLF in tank or barrel. A recent trend for high-end red wines has been to move them to barrel before MLF is fully complete, and then do MLF in barrel. Some winemakers, however, believe they get better results and more authentic wines by carrying out malolactic in tank before going to barrel.

The existence of frozen or freeze-dried cultures of *O. oeni* for starting MLF makes this choice a matter of winemaking style. Such frozen or freeze-dried cultures are added directly to the wine—in the past some cultures required a reconstitution stage—but they are faced with quite a task, because they are being added to a fairly hostile environment. These cultures of *O. oeni* are selected for the ability to tolerate low pH and high ethanol, but they aren't selected for their resistance to sulfur dioxide because this would then make it tricky to stabilize the wine at a later stage. Single strains are best because combinations of strains may end up competing with one another.

A recent shift has been experimentation with co-inoculation of cultured yeast and LAB. There are clear advantages of completing alcoholic and MLF simultaneously for commercial wines, including reduced time in tank or barrel and reduced risk of unwanted microbial activity.

However, the strains of yeast and bacteria need to be carefully matched in order for this to work.

THE TRANSFORMATION: MALIC TO LACTIC ACID

The transformation at the heart of MLF is that of malic acid ("malum" is Latin for apple), with its sharp, green, appley taste to lactic acid ("lac" is Latin for milk), which tastes softer and more appealing. The reaction involved is a decarboxylation, and it releases carbon dioxide. For this reason, wines that are undergoing MLF often have a bit of a prickle to them when you taste them, and if a wine is unstable and MLF starts in the bottle, then the wine will end up being slightly fizzy.

Malic acid is present in large quantities in unripe grapes. As grapes ripen, its concentrations diminish, and by harvest time it constitutes around one-tenth to one-quarter of total grape acids; the other two significant organic acids in wine are tartaric acid (the main one) and citric acid (present in lesser quantities). The exact amount will depend on the climate, and ranges from 2–6.5g/l (tartaric acid is found at 5–10g/l), although in grapes harvested from very cool climates malic acid levels of 15g/l have been reported. Malic acid, which has quite a sharp taste, is initially the main grape acid and can occur at 25g/l before *veraison* takes place, but this diminishes rapidly as the last stage of ripening takes place.

During MLF the LAB convert this malic acid to lactic acid, which is a softer-tasting acid and also a less powerful one. So, there is a pH shift and a loss of acidity with this second fermentation, the degree of which will depend on how much malic acid was present in the grapes in the first place. MLF happens in almost all red wines, and the decision whether or not to permit it to occur in whites is usually a style decision on the part of the winemaker.

But there is a lot more to MLF than simply this acidic conversion. Just as yeasts have a significant sensory impact above and beyond simply converting sugar to alcohol, LAB change the flavor of wine in ways that are only really now becoming appreciated as scientists take a closer look at the activity of these microbes. The sensory impact of LAB can be both positive and negative, largely depending on the strain of bacteria doing the fermentation, the presence of certain substrates in the wine, and the conditions under which the fermentation takes place.

The LAB feed off any sugars that remain in the wine after the yeasts have finished (yeasts leave a bit of hexose and pentose sugars) and grow in numbers. The bacteria need less than 1g/l of sugars to create a biomass sufficient enough to carry out MLF. Alongside this activity, they are also able to do the conversion of malic to lactic acid, but it is worth emphasizing that this is just one of their metabolic activities. During the course of their growth they are able to secrete a range of flavor compounds into the wine. This is worth exploring in more detail, because this is where bacteria have the potential for enhancing quality or impacting on it negatively. Indeed, it is really instructive to taste experimental wines inoculated with different strains of cultured LAB. This type of comparison shows the degree to which MLF can change the taste of the wine, in addition to modifying the acidity. But there is very little written on the subject, and most winemakers allow MLF to happen spontaneously, trusting that they are going to get a decent strain of LAB.

LAB produce acetic acid from metabolizing sugars, so this will increase VA. The degree to which VA is increased depends on how much sugar they metabolize, so this is a potential concern when MLF starts while the yeasts have not quite finished their job. Again, this VA increase is quite strain dependent. As well as degrading malic acid, some strains of LAB can degrade tartaric acid, too. Fortunately, only a very few strains can do this, but it is very bad when it happens. Pasteur called this the "tourney" disease.

THE FLAVOR AND OTHER EFFECTS OF MALOLACTIC FERMENTATION

One of the most well-known sensory impacts of LAB is the production of diacetyl (2,3-butanedione). This has an odor-detection threshold of 0.2mg/l in white wines, and 2.8mg/l in reds. It is formed by LAB from citric acid, which is one of the main organic acids in wine. Diacetyl has a distinctive buttery, creamy character, and at small levels can be attractive. But higher levels of diacetyl are not pleasant and can be considered a fault. And there are some circumstances where detectable diacetyl is undesirable. The factors favoring diacetyl production are the presence of oxygen, high concentrations of citric acid and sugar, temperatures below 18°C (64°F), and the removal of yeast cells before MLF. The levels can be reduced by the presence of viable yeast cells and the addition of sulfur dioxide (but not enough to inhibit all the bacteria). Diacetyl can react with cysteine (an amino acid containing sulfur) to produce thiazole, which smells of toast, popcorn, and hazelnut.

One of the most interesting contributions to wine flavor from LAB is the production of volatile sulfur compounds (VSCs). These are produced by the metabolism of the sulfur-containing amino acids cysteine and methionine, and the VSCs that result can be good or bad, depending on the context. These are the compounds implicated in reduction problems in wines: sulfides, disulfides, thioesters, and mercaptans (thiols). Acrolein, which is a bitter-tasting compound, is produced by some LAB strains by the degradation of glycerol. It is undesirable at any level, but fortunately only a few strains produce it.

One of the most talked about by-products of MLF is the formation of biogenic amines. All fermented products contain them, but LAB are capable of producing reasonably high levels. They are formed by the decarboxylation of amino acids, and the major ones found in wine are histamine, tyramine, putrescine, and phenylethylamine. They can have a range of effects on people sensitive to them, including headaches, breathing difficulties, hyper- or hypotension, allergic reactions, and palpitations. People differ in their sensitivity to them, but their presence in wine is undesirable. Not all strains of LAB are able to decarboxylate amino acids. The higher the wine pH, the more complex the range of bacterial species that will grow in it. As a result, there will usually be higher levels of biogenic amines. And white wines—which usually have a lower pH—tend to have lower levels of biogenic amines. Although sulfites are often blamed for allergic reactions to wine, it is much more likely to be the biogenic amines that are responsible, although this has not been proven conclusively. Using selected strains of LAB to inoculate for MLF is one way to reduce the risk of biogenic amines in wines. Currently, there are no regulations for biogenic amine levels in wine, but this could change.

But biogenic amines are not the worst thing that certain strains of bacteria produce. Ethyl carbamate is a carcinogen that is found in many foods and drinks, and it is formed through reactions between alcohol and a precursor such as citrulline, urea, or carbamyl phosphate. The main contributor to ethyl carbamate levels in wine is urea that is formed by yeasts from the degradation of arginine, but even after alcoholic fermentation some arginine (0.1–2.3g/l) remains in the wine, and LAB can produce citrulline as an intermediate in the degradation process of arginine. The USA has regulations for maximum levels of ethyl carbamate in wine, which they set at 15µg/l; Canada set a level of 30µg/l; the EU has no uniform maximum level. Typically, wine contains around 10µg/l while fortified wines contain around 60, but these levels can vary.

One positive effect of LABs is that some strains are thought to have glycosidase activity. Many of the flavor molecules in grape juice are in a chemical state where they need to be converted during fermentation to be active. A glycosidase is an enzyme that removes sugar groups, and in this case it can hydrolyze sugar-bound monoterpenes to release them as volatile aromatic monoterpenes. There is also some evidence that LABs are able to synthesize esters, which are fruity smelling compounds, but this needs to be verified further.

LABs are able to remove green flavors from wine. The reduction of vegetative/green-grassy aromas that can occur during MLF is thought to be through the metabolism of aldehydes such as hexenal, which contribute to these green flavors (along with methoxypyrazines). LABs are also thought to be able to improve the body of a wine, for example, through the production of polyols and polysaccharides.

Citric acid present in wine can be degraded by LAB, resulting in elevated levels of acetic acid (VA). It is degraded to pyruvic acid and then acetic acid, and this process also results in the production of diacetyl, butanediol, and acetoin. Diacetyl production is a major issue for MLF because of its pronounced sensory effects. It yields smells of butter, popcorn, and yogurt. At low levels (below 4mg/l) it can be quite nice in some wines, but at higher levels (typically 5mg/l and above) it isn't.

Of all the flavor impact of MLF, the main one is the change in acidity, which is quite significant. It usually increases pH (makes the wine less acidic) by 0.1–0.3 units and reduces TA by 1–3g/l.

The pH level of the wine makes quite a difference to the species/strain of LAB that get involved. Above pH 3.5 MLF occurs faster, and

Pediococcus and *Lactobacillus* are more likely to be involved. This can be a bad thing, because it increases the risk of cheesy, buttery, or milky off-flavors developing. High pH is also a huge risk factor for *Brettanomyces* in red wines, in part because sulfur dioxide is much less effective at higher pH, and because malolactic raises pH this is quite an issue. Lower pH (below 3.5) is desirable for MLF with lower risk of off-flavors developing. However, if the pH is really low, then MLF may struggle to get going.

WHICH WINES NEED MALOLACTIC FERMENTATION?

Almost all red wines need to complete malolactic fermentation. Wines made in warm climates may be made from grapes that have very low malic acid levels to start, so MLF makes relatively little difference. Many whites undergo malolactic fermentation, but here this is more of a style issue. Those who want to preserve freshness and acidity in their wines may choose to carry out partial MLF. For white wines, MLF is in part a stylistic choice. If winemakers want to prevent it from taking place, the usual method is to discourage the growth of bacteria by a number of interventions, including chilling, adding sulfur dioxide, racking the wine, and taking clarification measures. In addition, in many countries the use of the enzyme lysozyme, which specifically attacks bacteria, has been approved.

Fermenting in the vineyard

I meet Dom Maxwell on a Wednesday afternoon at Vagabond, a wine bar in London. He is over from New Zealand selling wine and meeting family. He is the winemaker for Greystone, located up in the head of North Canterbury's Waipara Valley, in the Omihi subregion. Maxwell has changed how he works over time. His first job was to make clean wines, so he inoculated. Gradually he has moved toward organics and natural ferments. The wine that Maxwell has become famous for is his vineyard ferment Pinot Noir. How did this come about?

"I was driving around the vineyard with Nick Gill, who I've worked with from the start. This is back in 2013, just before harvest, and I said I wouldn't mind fermenting something in the vineyard this year. He said, sounds good, what do you need to make it happen? Some pallets and access to a tractor. The idea was that we've been fermenting everything with natural yeast, but when we moved to Muddy Water [the sister property where they now make the wines] I noticed that our Pinot Noir ferments were different. Some people were using *pied de cuves* and there was a bit of chat locally about the need to get your vineyard yeast into the ferment. I was thinking about it, and though you'll still be bringing it into the winery and depending on what's happened in the winery in the past, how will you really know? We figured, it is worth a go. We fermented 700kg [1,543lb] the first year, and it took about twelve days to kick off. It was a cold part of vintage. Immediately, the first fork in the road popped up: it's not all about the yeast when you are fermenting out there, there's also the weather. We really liked the wine. It was quite dark because of all that cold soak, but we carried on. Every vintage has been different, and the weather has played quite a big part. There have been years when batches of fruit have been picked

at 20°C [68°F] and they have been fermenting by the next morning. Ordinarily, the trained winemaker in me would freak out. But when you look at those wines later on, they are beautiful. We taste them in the cellar and compare them with our traditionally made Pinot Noir, which has had a four-day cold soak and has been plunged twice a day, and we see the fineness and the detail of the vineyard ferments. There are still a lot of questions for us around the microbial population. We have just started a three-year research program with a Ph.D student from Lincoln University. She is in the middle of analyzing around three hundred samples we took over harvest, comparing the same fruit fermenting in the vineyard and the winery."

The grapes were initially processed in the winery and then taken out to the vineyards. But they have now purchased a single-phase destemmer and they can process everything in the vineyard. "We pick into buckets, they go onto the back of a quad bike, which drives up to the fermenter with the destemmer sitting on top of it hooked up to a generator." The goal of the research project is to look at the microbial population in the soils and vines, and how this relates to what is in the fermenter.

Is anyone else doing this now? "Yes. Alistair Maling came over to me at a tasting and showed me a photo of them at Te Kairanga [Martinborough] with a vineyard fermenter," says Maxwell. Helen Morrison has been doing some whole-bunch Sauvignon in the vineyard at Villa Maria in Marlborough. Amisfield in Central Otago are doing some, too. And someone who worked with Dom is making Sangiovese in the vineyard in Tuscany. "I have learned so much from this process," says Maxwell. As a result of what they have learned in the vineyard ferments, "we've gradually pulled our foot off in the winery, and our whole approach has changed."

9 Wine flavor chemistry

Wine is complicated. It is a bewildering mix of chemicals, some of which have flavor. And a crucial point here is this: flavor molecules have flavor because we can perceive them. It is therefore essential to understand the nature of human perception in order to make sense of wine chemistry. Typically, scientists have taken a reductionist approach to understanding the molecules contributing flavor to wine, breaking wine into its component parts. One by one, they have attempted to isolate flavor-active molecules and then looked to see how they smell. This is a useful start, but the overall flavor of wine is a result of the combination of lots of different odorants (molecules with a smell) and tastants (those with a taste), working in combination. They interact with each other and even molecules below their perception threshold can influence the overall "taste" of the wine.

What is the goal of wine flavor chemistry? Do we need to be able to put a chemical name to all the nuances of a fine wine, in order to be able to appreciate it? No, clearly not. But if we understand the precise mechanisms by which certain components of the grape must are transformed into beneficial flavor molecules—for example, by the metabolic action of yeast, barrel aging, or bottle maturation—then winemakers can adapt their techniques to maximize positive flavor development. In a similar vein, viticulturists can adapt their techniques to encourage the formation of precursors of positive flavor molecules and avoid the development of grape constituents that have a negative impact on wine characteristics. There is also a dark side to this

branch of wine science: a greater understanding of the impact of particular flavor compounds that occur naturally in wine will aid those who are inclined toward "creative" winemaking, because it would be very hard to detect the addition of flavor compounds to wine, particularly if these were added in tiny amounts (it doesn't take much of most volatile wine compounds to have a significant effect) or as aroma precursors. As well as being dishonest, undisclosed manipulation of this kind could have the potential to mar the "natural" image of wine in the eyes of consumers.

THE TASTE OF WINE: FACTORS INFLUENCING WINE FLAVOR

The factors contributing to the flavor of wine are still far from being fully understood. Here, I'm using the term "flavor" to refer to our perception of wine. People also talk about the "taste" of wine, but this is ambiguous, because they could be referring to either the specific information that comes from the sense of taste itself, or they could be referring to the global "taste" of wine where this term is used to describe all of what we sense when we drink wine.

The emerging understanding of flavor is that it involves the combination of a number of senses, and draws on input from the senses of taste (technically known as gustation, referring to the information that comes from taste buds in the tongue and mouth), smell (technically olfaction, with the information coming from the olfactory receptors in the nasal cavity), touch (the feel of the food or liquid in the mouth), and even vision (visual cues have been shown to shape the perception of flavor). In addition, there is input in the form of our context or previous experience—what we know about what we are about to taste can alter our perception. Thus flavor is a "multimodal" sensation, one in which all these different sensory inputs are combined together in the brain to give a unified perception of flavor, with perhaps the strongest influence being that of olfaction.

Any study of wine flavor has to consider both the physical properties of the wine and also the way that humans sense these. A wine has physical properties, which can be measured. These are "real," in that the wine has a chemical composition. If several researchers examine the wine using analytic tools, such as spectrophotometry or gas chromatography-mass spectrometry (GC-MS) we can expect them to get the same results. (Any differences in the measurements will be because of the calibration and accuracy of the tools they are using.) It follows that wine has a large number of aromatic compounds, which can be detected by the human olfactory system. It also has chemicals that elicit a taste response. But while this chemical composition is a property of the wine, the "taste" of a wine is not. It is a result of an interaction between the taster and the wine, with the taster bringing something to the encounter that significantly alters the wine's perception.

Humans do not "taste" in the same way that measuring devices such as pH meters work. Instead, the sensory information gained during tasting—encoded as electrical signals by the olfactory receptors, taste buds, touch receptors, and visual photoreceptors—is then subjected to some complicated processing by the brain, before an edited version of this reality (the conscious perception of the wine) is generated.

Our context and experience, as well as our expectations, shape this perception. Researchers have demonstrated that experienced wine tasters process sensory information related to the taste of wine quite differently from novice tasters. And supplementary information, such as the price of the wine or its quality level, also affects perception. So, what we call the taste of wine is dependent on brain processing. Because we all differ in our olfactory receptor repertoire, it is certain that we are all, to a degree, living in our own unique taste world (see page 204). It would therefore be more accurate to say that the taste of a wine is a property of the taster, but one that is based on the chemical composition of the wine.

This is not to say that wine tasting is totally subjective, and that ratings and tasting notes of wines are useless. Although taste is a property of the taster, there is a surprising amount of agreement about how wines taste. This is because much of wine appreciation is learned. When we taste the same wine together, we are sharing a good deal of common experience.[1]

THE CHEMICAL COMPOSITION OF WINE

Wine is complicated mix of many hundreds of flavor-active compounds. The exact number of volatile molecules found in wine is unknown but estimates generally fall in the range of 800–1,000. There are clearly many of them, although in each wine only a limited number of these are found above the level at which most humans would detect them, known as the perception threshold. But the complexity of wine flavor and aroma goes beyond just the large number of volatile compounds it contains. In particular, there are two factors that increase this complexity. First, people differ in their sensitivity to various taste and smell molecules. Some people lack the ability to sense specific taste or aroma compounds (the term for this deficit is "aguesia" with regard to taste and "anosmia" with regard to olfaction). As well as

---

1 For an extensive discussion of the perception of wine, see my 2016 book on the subject *I Taste Red: The Science of Tasting Wine* (University of California Press).

this, people have a range of different thresholds for compounds, meaning that some may be a little more sensitive to a specific smell than others. Second, wine aroma and flavor are not additive, in that these are simply a sum of the different smells and tastes of the various chemicals a wine contains. Instead, there are many interactions between the different components, including masking interactions (where one compound interferes with the perception of another) and synergistic interactions (where a perception is created by a combination of two or more different compounds).

Thus, there are two elements here that anyone studying the flavor of wine must grapple with. First, what is chemically present in the wine and, as many of the active compounds are potent and thus their presence at even tiny levels can be significant, this presents a challenge to analytical chemists. Second, we need to understand how this mixture of chemicals is actually perceived. Our final perception of a wine is the result of complicated processing in which all the different tastes and aromas of wine are integrated into a single perception, with us bringing quite a lot to the wine-tasting process. Vicente Ferreira of the Universidad Zaragoza in Spain is one of the leading experts working on wine composition and aroma. Ferreira sums this up by saying that wines don't have a single characteristic aroma, but "rather they have a palette of different aromas, which are difficult to define and which surely are perceived differently by different people."

Ferreira separates the various flavor compounds of wines into three different groups, which is a structured way of thinking that helps us grapple with this subject. In addition, there is the important concept of the wine matrix. While some wines contain what he terms "impact" compounds, many lack these, and instead contain a large number of active odorants, each adding nuances to the wine. One of the challenges for sensory scientists is that in many cases it simply isn't possible to establish a clear link between a sensory descriptor and a single aroma molecule. Instead, what tasters refer to by specific descriptors is often the result of the interaction of two or more odor-active chemicals. "When I started my Ph.D, wine aroma was about finding a molecule to explain everything," says Laura Nicolau, one of the researchers involved in the New Zealand Sauvignon Blanc program (see page 133). "By the end, people started to think it is not only one, it should be a combination. Vicente Ferreira was the first, to my knowledge, to talk about this."

WINE ODOR

Ferreira describes the basal composition of what he refers to as "wine aroma", which is the result of 20 different aromatic chemicals that are present in all wines to make a global wine odor. Of these 20 aromas, just one is present in grapes (β-damascenone); the rest are produced by the metabolism of yeasts, in many cases working on precursors present in the grape juice. These are:

- Higher alcohols (e.g. butyric, isoamylic, hexylic, phenylethylic)
- Acids (acetic, butyric, hexanoic, octanoic, isovaerianic)
- Ethyl esters from fatty acids
- Acetates and compounds, such as diacetyl
- Ethanol.

The influence of alcohol (ethanol) is quite strong. Ethanol has been shown to modify the solubility of many of the aroma compounds, and makes them more reticent to leave the solution, thus making the wine less aromatic. A study by Whiton and Zoecklein from 2000 showed that as alcohol rose from 11 to 14%, there was reduced recovery of typical wine volatile compounds in an analytical chemistry experiment. In 2007, Ferreira's group identified a range of esters responsible for the fruity berry flavors in a series of red wines. But when they added more of these to the wine, it didn't increase the fruity impact, because of the suppressing effect of other wine components, including alcohol. They showed this in another experiment in which they added increasing levels of ethanol to a solution of nine esters at the same concentration they are found in wine, and discovered that the fruity scent quickly falls as alcohol rises, to the point that when alcohol reached 14.5%, the fruity aroma had been totally masked by the alcohol.

CONTRIBUTORY COMPOUNDS

There are also another 16 compounds present in most wines, but at relatively low levels, which here I have labeled as "contributory compounds." Their odor activity value (OAV; the ratio of the concentration of the compound to its perception threshold) is usually below one, but they have odor activity that is synergistic, contributing to characteristic scents despite being at lower concentrations that would normally lead to them being smelled. These include:

- Volatile phenols (guiaicol, eugenol, isoeugenol, 2,6-dimethoxyphenol, allyl-2,6-dimethoxyphenol)
- Ethyl esters
- Fatty acids
- Acetates of higher alcohols
- Ethyl esters of branched fatty acids
- Aliphatic aldehydes with 8, 9, or 10 carbon atoms
- Branched aldehydes, such as 2-methylpropanol, 2-methylbutanol, 3-methylbutanol, ketones, aliphatic γ-lactones
- Vanillin and its derivatives.

IMPACT COMPOUNDS

Impact compounds are a group of chemicals responsible for the characteristic aromas of certain wines, even if present at very low concentrations. These are of great interest because they often contribute to distinctive varietal aromas. However, many wines lack distinct impact compounds.

For example, Sauvignon Blanc is very interesting as a grape variety because much of its characteristic aroma is believed to come from a small number of impact compounds, chiefly methoxypyrazines (of which the most significant is 2-methoxy-3-isobutylpyrazine), and three thiols (4-mercapto-4-methylpentan-2-one [4MMP], 3-mercaptohexan-1-ol [3MH], and 3-mercaptohexyl acetate [3MHA]). These have therefore been the focus of intensive research

Examples of impact aromas

Methoxypyrazine: the most important one is 2-methoxy-3-isobutylpyrazine (MIBP; known widely as isobutyl methoxypyrazine), which has a detection threshold of 2ng/l in water and white wine (slightly higher in reds), and is responsible for green, grassy, and green-pepper aromas; 2-isopropyl-3-methoxypyrazine (isopropyl methoxypyrazine) is also important, but likely secondary to MIBP. The methoxypyrazines are one of the few classes of impact compounds formed in the grapes and are highly stable through fermentation and aging.
Monoterpenes: such as linalool, which is important in many white wines, such as Muscat, and has floral, citric aromas.
Rose-cis oxide: characteristic of Gewürztraminer, this has a sweet, flowery, rose-petal aroma.
Rotundone: a sesquiterpene that gives pepperiness to Syrah, at incredibly tiny concentrations. Remarkably, one-fifth of people cannot smell this.
Polyfunctional thiols (mercaptans): These include 4MMP, which has a box-tree aroma (4.2ng/l detection threshold), 3MHA, which has a tropical-fruit scent (60ng/l), and 3MH. These three are important in the aroma of Sauvignon Blanc.

THE NONVOLATILE WINE MATRIX

But in addition to the actual aroma molecules, some of Ferreira's most interesting work has been on what is called the nonvolatile matrix of wine. The idea here is that wine constituents that don't have any aromatic characteristic of their own strongly influence the way that the various aromatic molecules present in wine are perceived. In effect, the nonvolatile matrix influences how we interpret the smell of wine.

Ferreira and his colleagues recently reported an interesting experiment in which they showed that the nonvolatile matrix is critical in determining the aromatic character of wine, even to the point that when the aromatics from a white wine are put into a red-wine matrix, the wine smells like a red wine.

"Knowledge of volatile and nonvolatile composition alone is not enough to completely understand the overall wine aroma and in general its flavor," state the authors in the introduction. "Interactions among odorants, perceptual interactions between sense modalities, and interactions between the odorant and different elements of the wine nonvolatile matrix can all affect the odorant volatility, flavor release, and overall perceived flavor or aroma intensity and quality."

For this study, they selected six different Spanish wines, all of which are available commercially (from Somontano producer Viñas del Vero):

1. Aromatically intense unoaked Chardonnay (fruity white)
2. Barrel-fermented Chardonnay (protein-rich white)
3. A young, light Tempranillo (neutral red)
4. A four-year-old barrel-aged 90% Tempranillo/10% Cabernet Sauvignon blend (highly structured polyphenol-rich red)
5. A three-year-old Tempranillo with marked astringency (very astringent wine)
6. A three-year-old Tempranillo with marked woodiness (typical woody aroma).

The aromatics from samples of each of these wines were removed by the use of a process called lyophilization (freeze-drying) and then any remaining aromatics were removed by using a chemical called dichloromethane. The dichloromethane was itself removed by passing nitrogen through the sample until it was all gone. The extract was then dissolved in mineral water to produce the wine matrix.

In a separate series of manipulations, aromatic extracts were collected from each of the wines, producing six aroma extracts. Then combining different wine matrices with different aroma extracts made a series of reconstituted wines. 20ml (0.7fl oz) of aroma extract and 120ml (4fl oz) of nonvolatile extract were combined, both equivalent to 600ml (20fl oz) of wine, along with 52ml (1.75fl oz) of ethanol to bring the alcohol level up to 12%, and enough mineral water to bring the sample to 600ml (20fl oz) in all. In total, 18 reconstituted wines were made, which a trained sensory panel then analyzed.

The results showed that the nonvolatile extract (the matrix) had a surprisingly large impact on aroma perception of the wine. As an example, when the aroma extract from wine 1 (fruity white) was reconstituted with the nonvolatile extract from the other white, there was relatively little effect. But when it was combined with the red wine nonvolatile matrix, there was a large difference, and the sensory panel started using terms relating to red fruits, rather than terms typically used to describe white wines. Other red-wine terms that started appearing were spicy and woody.

The effect was most marked when the nonvolatile matrix came from the astringent red wine, number 5. But a similar effect occurred when red-wine volatiles were added to a white-wine matrix, and white, yellow, and tropical fruits all start appearing in judges' tasting notes.

These results surprised the authors. Previous studies have shown that nonvolatile components of wine can affect wine aroma, but this is largely through binding to them and making them less releasable. The remarkable thing about this study is that it demonstrates that the nonvolatile matrix is having an important affect in actually modifying the perception of the volatile components of wine. It is well known that cross-modal sensory effects can modify perception, especially when vision is involved (experts describe white wines colored red using red-wine terms), but this was avoided in this study by using black glasses and asking participants to describe the aromas before they tasted the wine. The authors also demonstrated that the nonvolatile matrices were free of traces of aromatic chemicals. They found just tiny quantities of a few of the most polar odorants, but at quantities well below their perception thresholds.

What is emerging from these types of studies is a more holistic view of wine. While the reductionist approach attempts to study wine flavor by breaking it down into its constituent chemical compounds and then studying these in isolation, the field of wine-flavor chemistry is maturing with the realization of the limits of reductionism. The reductionist approach has been incredibly useful, but a view of the flavor of wine that treats the wine as a whole, and takes the human side of perception into account, is likely to lead to a more complete and satisfying understanding of wine flavor.

Maurizio Ugliano, a researcher at the Università di Verona, agrees with this this sentiment. "I did a lot of work with fermentation and nutrients when I was in Australia," he said. "People always had a tendency to simplify these stories. For example, they'd say, these esters have been increasing with fermentation, and the threshold of these esters if we don't add nitrogen was below the threshold, and when we add nitrogen it is above threshold. The approach has been to isolate each compound from its context and talk about it individually, saying the effect of this variable is important on this compound because it is becoming above threshold and before it was below threshold. The reality is that when you have some manipulation that affects many things at the same time you can't study one compound in isolation and talk about an individual threshold." He thinks this is a problem with the current view of wine-flavor chemistry. "We currently don't look at these changes in a holistic matrix way." He adds, "aroma chemistry, from an analytical point of view, has been very 'omic,' since the beginning. People always took the approach that they needed to analyze as many compounds as possible in one single analysis. There is no point in analyzing the way that one ester changes over time. But from a sensory point of view, we struggle to introduce the concept of systems in the way we approach the changes of aroma compounds. It is difficult when you have all these combinations of things that you need to test in sensory work."

MORE ON THE DIFFERENT WINE FLAVOR COMPOUNDS

This is not the right context for a detailed discussion of the complex chemistry of wine flavor, so I will spare you the chemical structures and most of the long names—these can be found elsewhere. For the purposes of this chapter, it will be useful to take an overview of the key classes of wine flavor compounds, highlighting a few that are of particular interest. These compounds can be divided neatly into five groups: acids, alcohols, sugars, polyphenols, and volatile compounds, although there is some overlap.

ACIDS

Acid is a vital component of wine, helping to make it taste fresh, but also helping to preserve it. White wines with higher acidity usually age better than those with low. Red wines can get by with a little less acidity because they contain phenolic compounds that help preserve them.

The main organic acids found in grapes are tartaric, malic, and citric. Tartaric acid is the key grape acid and can reach levels of 15g/l in unripe grapes. It's quite a strong acid and is specific to grapes. In musts it is found in the range of roughly 3–6g/l. Malic acid is abundant in green apples and, unlike tartaric acid, is widely found in nature. Before *veraison* it can hit levels of 20g/l in grapes. In warm climates, it is found in musts in the range of 1–2g/l, and in cooler climates it occurs at 2–6g/l. Citric acid is also widespread in nature and is found in grapes at 0.5–1g/l. Other organic acids present in grapes include D-gluconic acid, mucic acid, coumaric acid, and coumaryl tartaric acid. Further acids are produced during fermentation, such as succinic, lactic, and acetic acids. In addition, ascorbic acid may be added during winemaking as an antioxidant.

This is the part where it gets quite confusing. There is no single measurement for acidity in wine. There are two measures, both with the initials "TA" (total and titratable acidity, see below), but which are different. And there is pH. And also volatile acidity (VA, largely acetic acid), but we are not going to consider this here, because it is smelled rather than tasted.

Let's begin with pH

It refers to the concentration of hydrogen ions (known as protons) in a solution. It's expressed as a negative logarithmic value, which means the lower the number the higher the acidity. And it also means that a solution at pH 3 has ten times more acidity (defined as protons) than one at pH 4 (corresponding to approximately the range of pH values found in wine, though it can sometimes drop a little lower than 3). This is where we need to get a bit technical. The "acidity" of an acid depends on something known as its dissociation constant, or pK_a. The lower the pK_a, the more dissociated the acid is, which means it releases more protons into solution. Sulfuric acid has a pK_a of around 1, so it is almost completely dissociated, making it a very strong acid (in terms of protons in solution). Of the organic acids, tartaric has a pK_a of 3.01, which means it is pretty strong. Malic is 3.46, lactic is 3.81, and carbonic acid is 6.52 (which means it has very little dissociation and is thus a weak acid).

If malolactic fermentation (MLF) takes place, then the malic acid will be largely converted to lactic acid by the action of lactic acid bacteria. Lactic acid tastes less acid than malic acid, contributing just one proton per molecule while malic contributes two. As a result, the pH of the wine shifts upward through MLF by 0.1–0.3 units.

Musts and wines are known as acidobasic buffer solutions. This means you have to work quite hard to change their pH levels. If you add acid to water, you can shift its pH very quickly, because there is none of this buffering effect. But the presence of other compounds in musts and wines makes it less easy to shift the pH, and it is a little easier to shift pH in wine than must. It is actually tricky to predict the pH of the final wine by looking at the pH of the must, because several things occur during the winemaking process that can change pH. Where acidification is needed, it is usually done with tartaric acid, and as a rule of thumb, 0.5–1g/l of tartaric acid is needed to shift pH by 0.1 units. Legally, you could change pH with malic or citric acid, but because these are weaker acids, it would require quite a lot more. And adding citric acid isn't a great idea where MLF is going to take place, because the bacteria turn citric acid into diacetyl, which has a buttery taste and can be unpleasant. However, I know of some winemakers who use malic acid to make small changes in pH because it doesn't fall out of solution in the same way that tartaric acid tends to, especially when there is potassium in the must or wine. Some winemakers in warmer climates have illegally used sulfuric acid to change pH, because it is very effective.

Typical pH levels for a white wine would be 3–3.3, while for reds they would be 3.3–3.7. However, I had a New Zealand Riesling with a pH of 2.65, and a while back a South African red that was delicious (and had aged well) despite a pH of 4.1. Those are extremes, though.

High pH isn't necessarily a bad thing. It can confer on a wine a deliciously smooth mouthfeel (think of some Provençale rosés or Northern Rhône whites, for example). Generally, though, winemaking at lower pH levels is safer because of the reduced risk of oxidation and microbial spoilage; pH affects the amount of sulfur dioxide that is present in the active molecular form. At pH 3, 6 percent of sulfur dioxide is in the molecular form, whereas at pH 3.5 only 2 percent is. If the wine gets up to pH 4, then 0.6 percent of sulfur dioxide is in the molecular form, and so a lot would have to be added for it to have any significant influence in protecting the wine. One famous New Zealand boutique winery used to be known for its rather interventionist red winemaking, acidifying to low pH and then, before bottling, de-acidifying to get the desired pH. This reduces *Brettanomyces* risk considerably, and helps in other ways, such as fixing color.

So, what about TA?

This stands for both total and titratable acidity. Total acidity is the total amount of organic acids in the wine. Titratable acidity looks at the ability of the acid in the wine to neutralize a base (an alkaline substance), which is usually sodium hydroxide. The endpoint is typically pH 8.2 and is indicated by the change of color of a reagent such as bromophenol blue or phenolphtalein. Total acidity is the best measure to use, but it is hard to measure in practice, so titratable acidity is used as an approximation of this, but it is by definition always going to be a lower figure than the total acidity. So, when you see the TA of a wine given, you can assume it is the titratable acidity. The unit it is expressed in is g/l, but here's another potential source of confusion. Most countries use "tartaric acid equivalent," but in some European countries it is given in "sulfuric acid equivalent," which will be two-thirds of the value of tartaric acid equivalent.

When it comes to the taste of acidity, what is more important: pH or TA?

Most of the literature on this suggests that it is the TA that gives the taste of acidity, and so the figure that's important to look for is not pH but TA. The confounding factor here is that pH and TA are usually correlated so they are hard to separate, in that low pH wines usually have high TA. But you can get higher pH wines with high TA, and here the acid would taste quite sour. The different organic acids do seem to have different flavors. Tartaric is hard, malic is green, and lactic is softer with some sourness. I find that often where warm climate wines have their pH adjusted by tartaric acid, the levels of tartaric acid necessary can mean that the acid sticks out as very hard and angular, even where the pH is not especially low. Another issue is that added tartaric acid reduces potassium concentrations in the wine (they bind to form potassium bitartarate), and potassium is thought to play an important part in contributing to the weight or body of the wine.

SUGARS AND SWEETNESS

Sweetness in wine is a combination of three factors. First of all, there is sugar itself. This is sensed by sweet-taste receptors on the tongue. Second, there is a sweetness that comes from fruitiness. While "sweet" is tasted, some wines can also smell sweet, even though sweetness is a taste modality.

Most commercial red wines are dry in terms of sugar content, but many have sweet aromas from their fruitiness. Very ripe fruity flavors taste and smell sweet even in the absence of sugar. The third source of sweetness is alcohol itself, which tastes sweet.

It's really instructive to try the same red wine at different alcohol levels, where reverse osmosis or the spinning cone has removed the alcohol. As the alcohol level drops, with all other components remaining the same, the wine tastes drier and less rounded and full. Where alcohol has been reduced substantially, such as in the new breed of 5.5% alcohol wines, which are lighter, it's necessary to add back some sweetness, usually in the form of residual sugar. It helps if the starting wine had a very sweet fruit profile to begin with, too. For lower alcohol whites, blending in some Muscat or Gewürztraminer, which have sweet aromas, helps substantially.

There are a number of ways of making a wine with some residual sugar levels. For some white wines, fermentation stops naturally, or slows to a point where simply chilling and/or adding a little sulfur dioxide very easily stop it. It can, of course, be deliberately stopped in this way at any stage, but if fermentation is still ticking along nicely then more of both (chilling and sulfur dioxide addition) will be needed. Blending in must or grape juice concentrate to a dry wine can also make a sweet wine. For commercial wine styles where just a few grams per liter are needed to round off the wine, this is most easily done on the blending bench than by attempting to stop the fermentation at an exact point.

In sweeter white wines and also Champagnes, sugar and acid balance are vital. The two play against each other. Sweetness is countered by acidity, such that a sweet wine with low acid seems much sweeter (and often flabbier) than the same wine with high acidity. In Champagne, a typical dosage for Brut (dry) Champagne is 7–10g/l, which helps offset the acidity but doesn't make the Champagne taste sweet. Botrytized sweet wines are prized because, as well as concentrating sweetness and flavor, the shriveling process of noble rot concentrates the acid levels, and the great sweet wines of the world have very high sugar levels as well as high acidity.

The perception of sweetness is altered by the aroma. Although we can't smell "sweet"—it is a taste modality—we have learned to associate certain smells with sweetness. Thus, we rate the sweetness of a sugar solution higher if it is presented with a sweet smell, and lower if it is presented alongside a savory smell. This could explain the phenomenon that as sweet wines age, they taste less sweet. If you get a 30-year-old Sauternes, it isn't nearly as sweet tasting as a younger wine, even if both started out with the same level of sugar to start with. This disappearing sweetness is described as

the wine "eating sugar." As far as we know, the level of sugar has not changed with time, but what could be occurring is the aroma of the wine, and its effect on the perception of sweetness. A young Sauternes, for example, usually has very high levels of fruity aromas, typically apricot, passion fruit, and peach. With time in bottle, these fruity aromas diminish. Smell an old Sauternes: it smells much less fruity than a young one, and usually has complex savory aromas. The brain is computing the sensation of sweetness not only from the sugar level, but also from the smell of the wine. Even if the old wine still has the same level of sugar, the perception we have of its sweetness has changed.

Support for this idea comes from a couple of recent studies on tomatoes and strawberries, carried out by Linda Bartoshuk and her colleagues at the University of Florida. They looked at the composition of a range of tomato varieties, testing the levels of sugar and also a group of volatile compounds. They then got a sensory panel to taste these tomatoes, rating them for a range of attributes, including sweetness. They then looked at which compounds contributed to this perception of sweetness: it turned out not to not only be sugar, but also a group of seven volatiles. For example, one variety had 45g/l of sugar and was given a score of 13 on the perceived sweetness scale, while another had less sugar (just under 40g/l) but got a score of 25. It got this big score because it had about twice the level of a group of six volatiles that were correlated strongly with sweetness. In the strawberry work, the researchers found 24 volatile compounds that showed significant correlations with perceived sweetness intensity, independent of glucose or fructose levels, and 20 that did this independent of sucrose concentration. Of these, six altered sweetness perception independent of all three sugars.

What is happening here? There are two ways in which we smell something. The first is through the nose: orthonasal olfaction. The second is by aromas entering the nose around the back, from the mouth: retronasal olfaction. The brain deals with these two types of smell differently. "Taste" is computed from the taste sensations plus the retronasal olfaction, which are processed together. Sweet smells are likely to result from previous pairings of volatiles and sweet tastes, such as the smell of strawberries, peaches, vanilla, and caramel. So, as a sweet wine ages its smell changes. And even if the sugar level stays the same, the perception of sweetness will change.

ALCOHOLS

Aside from water, ethyl alcohol is the most important component of wine, and is produced by fermentation of sugars by yeasts. On its own, it doesn't taste like much, but the concentration of alcohol in the final wine has a marked effect on its sensory qualities. This is evidenced by the "sweet-spot" tastings carried out during alcohol reduction trials. If a wine with a high natural alcohol level is subjected to alcohol reduction via reverse osmosis, a series of samples of the same wine can be prepared differing only in alcohol levels, say at half degree intervals from 12 to 18% alcohol. Panels of tasters show marked preferences for some of these wines over others, and different descriptors are commonly used to describe the sensory properties of the different samples. Excessive alcohol can lead to bitterness and astringency in a wine. It may also taste "hot."

Meillon and colleagues conducted an interesting study in 2010. They took an Australian Syrah and reduced it in alcohol from its original strength (13.4%) down to 8%, with three wines at intermediate levels between these extremes. They showed these wines to 71 French consumers who drank red wines at least once a month, measuring their liking and perceived complexity. The Syrah at 8% was liked significantly less, but for the other wines there was no significant difference. Adding sugar to the 8% Syrah increased its likeability considerably. They found they could segment this population by their reaction to the wines. Group 1 (18 individuals) liked the 11.5% wine the best and liked the lowest two alcohol wines much less than the other groups. Group 2 significantly disliked the Syrah at 8 and 11.5%, but they liked the 13.5 and the sugared 8% wines. Group 3 preferred the two lowest-alcohol wines. The more consumers dislike the alcohol-reduced wine, the more bottles they have in their cellar—an interesting finding.

POLYPHENOLS

Polyphenols are so important in wine that they get their own chapter (Chapter 10, see page 140). They are probably the most important flavor chemicals in red wines but are of much less importance in whites. Polyphenols are a large group of compounds that use phenol as the basic building block. An important property of phenolic compounds is that they associate spontaneously with a wide range of compounds, such as proteins and other phenolics, by means of a range of noncovalent forces (for example, hydrogen bonding and hydrophobic effects).

VOLATILE COMPOUNDS

This is where things get really complex, but it is also where much of the action is. It is the volatile compounds that give wine its smell, referred to more respectfully as bouquet or aroma. Volatile compounds come directly from the grapes themselves, but more commonly are secondary aromas arising from fermentation processes, or even tertiary aromas developing during maturation and aging of wine. Most are present in extremely low concentrations, which, before the advent of highly sensitive analytical techniques, made their study a difficult business.

Rather than list the 400 or so thought to be important in wine, here follows a description of the main classes, with one or two specific examples.

Esters are especially important to wine flavor. They are formed by the reaction of organic acids with alcohols and are formed during both fermentation and aging. Ethyl acetate (also known as ethyl ethanoate) is the most common ester in wine, formed by the combination of acetic acid and ethanol. Most esters have a distinctly fruity aroma, with some also possessing oily, herbaceous, buttery, and nutty nuances.

Although aldehydes are present in grape must, they are of relatively minor importance in wine flavor, with the exception of acetaldehyde (ethanal), which is a component of some sherries. Vanillin (4-hydroxy-methoxy-benzaldehyde) can be an important aroma molecule in wines aged or fermented in oak barrels.

Ketones include diacetyl (butane-2,3-dione), which gives a buttery odor at higher levels; this can be negative at higher levels. Acetoin (3-hydroxybutan-2-one) has a slightly milky odor. β-damascenone and α- and β-ionone are known as the complex ketones, or isoprenoids. The former has a roselike aroma and is most commonly found in Chardonnay. The latter occur in Riesling grapes and are said to smell like violets. They are also present in other wine types. Benzoic aldehydes are taint compounds with a bitter almond flavor that are sometimes produced as a result of the incorrect application of epoxy resin vat linings.

Some 40 higher alcohols, also known as fusel oils, have been described in wine. The most important of these are the amyl alcohols. With their pungent odors they are negative at higher levels but kept in check they can be positive. For example, hexanol has a grassy flavor.

Lactones (furanones) have been identified both in grapes and oak barrels during wine aging. The oak lactones (cis- and trans-β-methyl-γ-octalactone) are particularly important in barrel-aged wines, imparting sweet and spicy coconut aromas, together with woody characteristics. Sotolon (3-hydroxy-4,5-dimethyl-5(H)-furan-2-one) is a lactone associated with botrytized wines and also oxidation, and has sweet, spicy, toasty, nutty aromas.

The most significant volatile acid in wine is acetic acid, producing during fermentation but more significantly a result of *Acetobacter* activity. It tastes sour and smells like vinegar.

Volatile phenols are important in wine aroma: 4-ethylphenol and 4-ethylguaiacol, found predominantly in red wines, are formed by the action of the spoilage yeast *Brettanomyces*, and have distinctive gamey, spicy, animally aromas; 4-vinylphenol and 4-vinylguaiacol are rare in red wines and more common in whites, and also have largely negative aromatic properties. These are formed by the enzymatic decarboxylation of cinnamic acids, a process inhibited by some grape phenols in red wines.

Terpenes are a large family of compounds that are widespread in plants. Grapes contain varying amounts, and these survive vinification to contribute to wine odor. More than 40 have been identified in grapes, but only half a dozen are thought to contribute to wine aroma. They are highest in Muscat wines, and the distinctive floral, grapey character is down to the likes of linalool and geraniol. Other varieties, such as Gewürztraminer and Pinot Gris, also have a terpene component to their aromas.

Methoxypyrazines are important in the aroma of certain wines. They are heterocyclic compounds that contain nitrogen and are formed by the metabolism of amino acids; 2-methoxy-3-isobutyl-pyrazine is a distinctive element of the aroma of varieties such as Cabernet Sauvignon, Cabernet Franc, and Sauvignon Blanc. At higher concentrations this can be excessively herbaceous, and it is generally seen as a problem in red wines, but an asset in certain styles of white when it contributes less of the green pepper and more of the fresh, grassy type of aromas. Methoxypyrazines have extremely low detection thresholds (see the box, below, on "Green in wine").

Volatile sulfur compounds are important in wine aroma. Mercaptans (thiols) are negative at higher amounts, but in controlled quantities they are important in the aroma of wines made from Sauvignon Blanc and some other white varieties. Some sulfur compounds have positive effects on wine aroma at extremely low levels.

FLAVOR CHEMISTRY IN ACTION: THE KEY FLAVOR COMPOUNDS IN SAUVIGNON BLANC

A great example of wine flavor chemistry in action comes from the New Zealand Sauvignon Blanc research program. One of the objectives of the program was to characterize the aroma and flavor compounds present in Sauvignon Blanc, with a view to identifying why New Zealand (and specifically Marlborough) Sauvignon Blanc is so distinctive. Exactly how is it different? Is it because it possesses aromas and tastes that other Sauvignons lack? Or is it because it has particularly high levels of some compounds that it shares in common with Sauvignons from other regions?

Researcher Laura Nicolau went to New Zealand in 2003 to head up the wine aroma research side of the Sauvignon program. She had previously been working at the Université de Bordeaux, where analytic capability for some of the important aroma compounds had already been developed. This is where Tominaga and Dubourdieu had carried out the pioneering work on Sauvignon Blanc aroma. Together with Frank Benkwitz, her Ph.D student, she set about answering some of these questions using the techniques of analytic chemistry. Benkwitz's Ph.D thesis contains some extremely important results, which shed new light on the significant flavor compounds present in Sauvignon Blanc.

Carrying out this sort of research is not straightforward. While it is possible to identify a list of aroma and flavor compounds, calculate their concentration in wine, and then highlight those that are above perception threshold, wine flavor is not simply additive. The wine is a "whole," and these flavor compounds interact in complex ways to produce the overall flavor of wine. "I teach wine aroma to students, and we used to say that compounds that have lower odor activity values—where the concentration is under the perception threshold in the wine—are not so important," says Nicolau. "But the more research we do, the more we see that they are important, able to influence the perception of other components in the mixture."

A "MODEL" SAUVIGNON BLANC

Nicolau and Benkwitz therefore adopted a clever two-pronged strategy. In the first part of his work Benkwitz aimed to provide a list of Sauvignon compounds in order of estimated importance for overall aroma. To do this, he used a range of analytical techniques, including GC-O (gas chromatography–olfactometry), AEDA (aroma extract dilution analysis, a quantitative GC-O technique), and GC-MS (gas chromatography–mass spectrometry). Those compounds present at concentrations above their threshold detection level were highlighted as significant.

In the next stage he created a "model" Sauvignon Blanc by deodorizing an actual Sauvignon Blanc wine, and then adding back the key aroma-active molecules at the concentrations they were found in the original wine in a reconstitution experiment. Using this model wine, he was able to do omission tests, looking at the effects of omitting either related groups of compounds, or single compounds on their own, and seeing their effect on the perception of the wine by trained sensory panels. This is a particularly elegant experiment with great explanatory power. Significantly, it treats the wine as a whole, getting away from the limits of more reductionist approaches.

Benkwitz looked at 83 different examples of Sauvignon Blanc in all, the majority of which were from New Zealand, but also including typical examples from other countries. Initially the experiments were just with Marlborough Sauvignon and some Sauvignons from Bordeaux, but those from further afield later joined the samples.

In total, some 49 different aroma compounds were identified, with many only present in trace amounts. However, this initial trawl showed that all the Sauvignons are qualitatively similar, that is, they all contain the same compounds but at different levels. There are no compounds that are unique to the distinctive Marlborough style.

However, Marlborough Sauvignon Blanc is clearly different from most other Sauvignons in a number of respects. First, Marlborough Sauvignon shows quite high levels of methoxypyrazines. These are a group of compounds including 2-methoxy-3-isobutylpyrazine (MIBP; known widely as isobutyl methoxypyrazine), 2-methoxy-3-isopropylpyrazine (MIPP; known as isopropyl methoxypyrazine), and 2-methoxy-3-secbutylpyrazine (MSBP; known as sec-butyl methoxypyrazine). Of these, MIBP is the key player. While other Sauvignon Blanc examples also share high levels of methoxypyrazine, the Marlborough examples were consistently quite high in this respect. Second, a group of compounds known as polyfunctional thiols are present at unusually high levels in Marlborough Sauvignon.

Three are considered to be particularly important in Sauvignon: 3-mercaptohexanol (3MH), 3-mercaptohexyl acetate (3MHA), and 4-mercapto-4-methylpentan-2-one (4MMP). Of these, 3MH and 3MHA are found at extremely high levels in Marlborough Sauvignon. There is a striking variation within the region and from year to year, however, but on average Marlborough Sauvignon has much higher average levels than wines from other regions.

The purpose of this stage of the research was to produce a list of compounds with an odor activity value (OAV) of greater than one. Initially, a quantitative form of GC-O, AEDA, was used to identify target compounds. AEDA relies on calculating the maximum dilution at which a compound can be detected, by recording the flavor dilution (FD) factor. The FD value produces a hierarchical list of odorants ranked in terms of importance. This can be used as a screening method before the OAVs are defined, a step that takes more time to carry out.

Working out the concentration of an odorant in the wine under scrutiny, and then dividing this by the perception threshold calculates the OAV. A compound with an OAV of 1 is present at its perception threshold, and one with an OAV of 2 is present at twice its perception threshold.

The first stage in the reconstitution experiments was to create a deodorized wine. The wine chosen for this study was a Saint Clair Premium Marlborough Sauvignon Blanc from 2007, which included fruit from the three most important Marlborough subregions and had relatively high levels of 3MH (9,250ng/l) and 3MHA (1,350ng/l). Benkwitz and Nicolau used 5g (0.18oz) of a resin called XAD-4 to remove all the flavor compounds from 200ml (7fl oz) of wine. This resin absorbs small hydrophobic (water-hating) compounds, but it also had the side effect of reducing the acidity and alcohol a little, as well as taking out some polyphenols.

The next stage was to add back the aroma/flavor compounds at their original levels. Initially, they used a "complete" model with 19 compounds added back, but later a simpler model with just the 11 compounds having an OAV of more than 2 was considered sufficient. The model wine that they created was significantly different to the original wine, but Benkwitz speculates that it might be closer if the differences in pH, alcohol, and polyphenols were corrected.

The 19 compounds, in order of OAV, were: 3MHA, 3MH, isoamyl acetate, ethyl hexanoate, ethyl butanoate, hexanoic acid, 4MMP, MIBP, octanoic acid, ethyl octanoate, isoamyl alcohol, phenylethanol, hexyl acetate, 1-hexanol, β-phenylethyl acetate, 3-hexenol acetate, 3-hexenol, linalool, and α-terpineol.

WHICH MOLECULES REALLY MATTER?

Nicolau explains the idea behind the model wine experiments. "You deodorize the Sauvignon Blanc and put back the compounds at the level that you know they were present in the wine. Then you omit some of them in different combinations. We might look at a group of compounds, such as taking out all the esters, or all the thiols, or taking one ester or thiol at a time." The science group at Plant and Food Research in Auckland carried out the sensory analysis of the reconstituted wines. "The surprise in Sauvignon came from the terpenes," she points out. "If you take them out, it makes a huge difference to the overall perception of the wine." The two terpenes linalool and α-terpineol had a huge impact when they were omitted. Apple lollipop, stone fruit, and tropical characters were all scored as less. Monoterpenes have ten carbon atoms in their backbone, and include linalool, nerol, geraniol, citranellol, and α-terpineol. They interact synergistically and have pleasant floral aromas, but individually they are usually below threshold level in Sauvignon Blanc. Thus, it is of interest that they are having an effect.

"The esters also have a huge influence," says Nicolau. More than 160 of them are found in wine, and they are produced during fermentation. They are described as being fruity and floral and are rapidly hydrolyzed in wine over the first year after bottling, with lower pH accelerating this hydrolysis. "They have a broad impact, so they can influence the fruity aromas generally, including the tropical fruit," says Nicolau. Removing the esters showed a small drop in intensity for most of the descriptors, as well as a large decrease in "passion-fruit skin/stalk" and "sweet, sweaty passion fruit," both of which were previously thought to be associated with thiols. "Previously we thought this was from the thiols, but the esters have that note as well," says Nicolau. She adds that, "when you take out the thiols, there is a more subtle difference than when you take out the terpenes, for example." Indeed, when the three thiols 3MH, 3MHA, and 4MMP are taken out, the effect on overall aroma is not as significant as would have been expected. The descriptors "flinty," "passion-fruit skin/stalk" are significantly less intense, while "capsicum" is more intense.

The C6 compounds seem to be very important. These include 1-hexanol, cis- and trans-3-hexenol, and cis and trans-2-hexenol, and they are described as having herbaceous/green/grassy aromas. Eliminating these reduces tropical characters as well as "passion-fruit skin/stalk."

The key methoxypyrazine MIBP is described as having aromas of capsicum, vegetation, and green, but its omission does not change the intensity of the capsicum character in the wine. The only significant change is the decrease in intensity for flinty. This is surprising, because methoxypyrazines are considered impact compounds in Sauvignon Blanc, but the results from this experiment do not support that hypothesis.

One really surprising result was that in reconstitution experiments, taking out a single compound could have more of an effect than taking out whole groups of related compounds, including that single compound. This is hard to explain. For example, if the three polyfunctional thiols (3MH, 3MHA, or 4MMP) are removed together, there is a relatively small change in the aroma profile. If 4MMP or 3MH alone are removed, there is a more significant difference.

In the initial study, β-damascenone, a norisoprenoid that smells like fruit/roses, was omitted. In a later study it was included, and Benkwitz found that it enhanced the perception of the thiols quite significantly, but it only had a minor impact when it was omitted alone. He thinks that β-damascenone may turn out to be a very important compound in Sauvignon Blanc. Norisoprenoids are 13-carbon compounds produced by the degradation of chemicals called carotenoids during the ripening process in fruit. As well as β-damascenone, α- and β-ionone are also important, and have a violet aroma.

Another single compound with a relatively large effect if left out is β-phenylacetate. While its OAV in the model wine was just more than 1, it has a significant impact if omitted, reducing the overall intensity of the aroma profile, while slightly increasing the scores for apple lollipop and banana lollipop. When ethyl hexanoate is removed, it also has a significant effect, increasing "banana lollipop" and "apple lollipop," as well as tropical and sweet, sweaty passion fruit. "Passion-fruit skin/stalk" and "flinty" are decreased.

The results from these reconstitution experiments are really interesting and show what a powerful technique this is for yielding understanding about the flavor of wine. There is a lot more work to be done with these techniques and, as the model wines become more sophisticated, the results obtained are likely to be more reliable, offering greater insight. For now, though, this work has shown that the flavor of Sauvignon Blanc is more complicated than the simple model involving the impact compounds methoxypyrazines and the polyfunctional thiols that were previously identified. However, while these results are somewhat revisionist, they do not negate the important work already done on Sauvignon's impact compounds. In particular, the impressive work on thiols was at the heart of the New Zealand Sauvignon program. The facts that (1) the two thiols, 3MH and 3MHA, are the compounds uniquely elevated in Marlborough Sauvignon Blanc, and (2) there is a high correlation between the use of the descriptors "tropical," "sweet, sweaty passion fruit," and "passion-fruit skin/stalk," and the concentration of 3MH and particularly 3MHA suggests that the thiols are indeed important in Sauvignon aroma.

Manufacturing wine: should it be allowed?

Wine is usually defined as "fermented grape juice as an alcoholic drink," but there is also a supplementary definition of a "fermented drink resembling this made from other fruits etc., as specified [elderberry wine; ginger wine]." Unless specified otherwise, therefore, the assumption is that wine is the product of fermented grape juice. But with fantastically rare exceptions, winemakers add things to wine. At the simplest level, this addition may be limited to a little sulfur dioxide. Most commonly, more things are added, including yeast-starter cultures, fining agents, tannins, acidity, oxygen, sugars, and flavor compounds via oak barrels or barrel substitutes. Who is to say which additives are acceptable and which are not? Is it more acceptable to add more of a component that is already present in the wine, than to add one that is not? Bear in mind also that there are viticultural and winemaking techniques that will ensure that more (or less) of a particular chemical component ends up in the final wine. Is manipulation acceptable when, as in this case, it is indirect, but not if it involves the direct addition or subtraction of a chemical from the must or wine? And are direct manipulations that are traditional, such as *chaptalization* (the adding of sugar to must), more acceptable than ones that are novel (such as removal of alcohol from a wine by reverse osmosis)? And what about the selective use of yeast strains known to enhance the presence of certain flavor molecules in the finished wine? These are difficult questions to answer with any degree of certainty.

Wine mimicry

Let's be theoretical for a moment. If it were possible to analyze a great wine, such as Chave's Hermitage from a great vintage, and replicate this accurately by chemical means, would this be an evil thing to do? It would certainly be wrong if the resultant liquid were passed off as Hermitage. But what if it were sold without such labeling deceit, at a low price. It would offer someone of limited means a great sensory experience for relatively little money. Whether or not this is desirable or wrong depends on your view of wine. Is wine more than just what is in the bottle? Or is the experience of wine confined wholly to the mental representation that results from our response to sniffing and slurping on the liquid we have in our glasses? To a degree, neurobiology can answer this question because our knowledge of and belief about the wine we are drinking helps shape the sensory representation we have. So, for me, if I drank a fake wine almost identical to Chave's Hermitage in chemical properties, it would be a lesser experience than knowing that I was drinking the real thing, with the cultural context that this brings. On another level, someone who neither knew nor cared about Chave or Hermitage would have a near-equal experience of both the real wine and the manufactured liquid. The tentative conclusion might be that for certain markets, a manufactured winelike beverage would be perfectly acceptable, although manufacturing fine wine is unlikely to prove a success unless you are prepared to be fraudulent at the same time.

Green in wine

Can you taste a color?

In some cases, it seems that you can. Green. It's a descriptor often used in tasting notes, and often it is pejorative, especially in the context of red wines. Think of some of the tasting terms that are used to describe greenness: herbal, grassy, vegetal, asparagus, bell pepper, and even "green tannin."

There are a number of chemicals present in wine that can contribute to these green flavors, but there is one group of compounds in particular that is strongly associated with green aromas and flavors in wine. This is the alkyl-methoxypyrazines, and in our exploration of greenness we will begin with them. They are present in green tissues of plants, and they were first discovered in green bell peppers in 1969.

Sauvignon Blanc is one grape variety where it seems that green is desirable. Good Sauvignon often has an element of greenness. Think of a New Zealand Sauvignon Blanc from the Marlborough region, and especially the Awatere Valley. Key descriptors include bell pepper (also known by many as capsicum or green pepper), tomato leaf, herbal, and even cut grass. These aromas and flavors have chiefly been attributed to methoxypyrazines.

With the resounding success of New Zealand Sauvignon, winemakers around the world began working to increase the levels of methoxypyrazine in their Sauvignons. Some South African winemakers famously took the illegal shortcut of adding methoxypyrazines to their wines directly, but they got caught. Now, we think that green alone is not enough, and fortunately the excessively green Sauvignons with little else going on are becoming rarer.

Green is also a frequent contributor to the aroma profile of the Bordeaux grape varieties Cabernet Sauvignon, Cabernet Franc, and Merlot. Carmenère, a Bordeaux variety that relocated to Chile, also has high levels of green. Good red Bordeaux often has green in its signature, although in some New World countries winemakers seem terrified of greenness. In most other well-known red varieties, methoxypyrazines are below detection levels in ripe grapes.

Isobutyl-methoxypyrazine (IBMP) is the most important methoxypyrazine in wine, and typical concentrations found in wines would be in the range of 5–30ng/l. That's significantly above the sensory threshold in water, which is just 2ng/l (the sensory threshold in wine, especially red wine, is higher at around 15ng/l). The other ones identified in wine are isopropyl-methoxypyrazine (IPMP) and sec-butyl-methoxypyrazine (SBMP), which are usually at much lower levels, but could still play a role contributing to these green flavors.

How do these green flavors get into wine?

Methoxypyrazine levels gradually increase in grapes as the berries grow, reaching a maximum just before *veraison*, and then they start to decrease as ripeness approaches. Unripe grapes typically have high levels of methoxypyrazines, which may be acting as an antifeedant, along with high tannin levels. The idea here is that the grapevine does not want its berries eaten by birds until the seeds are ready to be dispersed. Certainly, if you try an unripe grape, it tastes pretty disgusting: there is no sugar, very high acidity, grippy tannins, and excessive greenness.

Vines with grape leafroll virus often have delayed berry maturation and can make red wines with high levels of green in them. Often, winegrowers leave the grapes on the vine in an attempt to ripen out the greenness, but it is not as simple as this since extra hang time may not be effective in reducing methoxypyrazine levels beyond a certain point. If they are present, they can't just be ripened out.

A common viewpoint among viticulturists is that one way of reducing methoxypyrazine levels is to expose the grapes to the sun. So, they pluck the leaves in the fruit zone, thinking that the sunlight will degrade the methoxypyrazine. But there is very little evidence for this photodegradation actually occurring. Still, leaf removal seems to have some effect, especially when it is done early on. This might be because of reduced accumulation rather than increased degradation. In fact, there is evidence that if the vine is growing rapidly during this period of methoxypyrazine accumulation,

then more methoxypyrazine is accumulated, independent on the level of cluster shading. So, for red wines you really want vine growth to slow down just before the methoxypyrazine accumulation phase.

If methoxypyrazines are present in the grapes at harvest, because they are very stable compounds, they tend to persist through fermentation and then stay at the same level in the final wine. One other source of greenness in wine is from the use of stems in winemaking. This is covered in Chapter 12, see page 152.

Interestingly, methoxypyrazines are also produced by some insects. There are notable incidences where ladybugs present in grape bunches at harvest cause green taints in wine, because they are capable of producing heroic levels of methoxypyrazine. This can lead to the wine being ruined. Vineyards near soybean farms can be at particular risk, because the soybeans are harvested and then the ladybirds find a new home in the vineyard. You don't need many in a vat for this green taint to occur, so a vibrating sorting table is a must where this is potentially an issue.

So, what can you do if you want lower methoxypyrazine levels in your wine, but they are present in the harvested grapes? It is possible to mask their presence using oak, and it seems that more gentle pressing results in lower levels in white wines. Some methoxypyrazine is lost in settling and clarifying the must of white varieties before fermentation. And thermovinification also reduces levels. But, generally, if you want lower levels, you need to address this issue in the vineyard.

Other green contributors

There is more to greenness in wine than just methoxypyrazines. The other main contributors are a group of chemicals called green leaf volatiles. These are produced by a chemical reaction chain called the oxylipin pathway, and they are involved in plant defense. One of these compounds is called hexanol, which is described as green, herbaceous, grassy, woody, and fruity. Then there is cis-3-hexen-1-ol, which is green and grassy (and also acts as a semiochemical, altering predators of herbivores that dinner is ready), and trans-3-hexen-1-ol, which is green, earthy, and fatty.

Finally, there are green notes in wine that could come from the environment. Vineyards in proximity to both eucalyptus and pine trees can make wines that are influenced by aromatics from the leaves of these trees, which impart distinctive characters to the wine, including an element of green.

What about green tannins? Personally, I think this is just picture language, and that they don't actually exist. Tannins are sensed as astringent (this is the sense of touch), and sometimes also as bitter (the sense of taste). But I can't see a way for them to taste green. Perhaps people using this term are describing tannins with a certain mouthfeel as green because they co-occur with green flavors from slightly underripe grapes.

10 Phenolics

We are about to embark on a difficult journey. I thought I should warn you at the outset. The subject of phenolics is a very important one in wine, but it's also a fiendishly complicated topic. And to make matters worse, it's one where our understanding is incomplete. (However, if you are a student of wine, you should be used to this by now. There are so many gaps in our knowledge.) The reason phenolics is so important to understand is because we use terms like phenolics, tannin, and anthocyanins all the time when we talk about wines and so we need to know what we are referring to.

"Phenolics" refers to a large group of chemicals that use a structure called phenol as the basic building block. That is where the name comes from: "poly" phenols are where more than one phenol group is joined to another. They are probably the most important group of flavor chemicals in red wines but are of much less importance in whites. Tannins are the most famous and important of the polyphenols in wine, but there are others. This is one area where it can get quite confusing, because the term "tannin" is not referring to a group of structurally related chemicals. Instead, it describes the behavior of certain polyphenols. So, when we attempt to break down the various classes of phenolics by structure, we end up with a somewhat overlapping classification.

Let us begin with the phenol group. A phenol is a chemical that consists of what is known as a benzene ring (a structure consisting of six carbon atoms joined in a ring with a hydrogen atom attached to each, formula C_6H_6) with a hydroxyl group (-OH) substituted for one of the hydrogens, and so its chemical formula is C_6H_5OH. While phenol is an important industrial chemical, we are interested here in phenolics that are naturally synthesized by plants, built up of one or more phenol groups, modified in various ways and joined together. Plants act as chemical factories, and there are more than 8,000 different plant phenolic compounds, which serve a range of roles. Most of the phenolics we will discuss here are polyphenols, but wine also has simple phenols with just one ring, such as caffeic acid.

PHENOLICS IN THE GRAPE BERRY

The grape-berry skin has two types of cells, an outer layer of clear epidermal cells and then several layers (around six but it depends on the variety) of hypodermal cells. Next there are three different pulp tissues: the outer and inner mesocarp and the vascular tissue that separates them. Then we have the seed, which has two layers of seed coat (testa) cells separated by a thin-walled parenchymal layer. The outer layer is a cuticle, and then between this and the lignified inner testa cells there is the layer of thin-walled parenchyma cells that contains almost all of the seed phenolics. Interestingly, seed browning is now thought to be because of the tannins present in this layer getting oxidized and is not associated with the process of lignification.

NONFLAVONOID POLYPHENOLS: THE PULP PHENOLICS

The pulp of the grape seems like a good place to start, because the polyphenols found here will be present in both red and white grapes and will be extracted even where grapes are pressed before fermentation, as is common with white wines. The first distinction we will make is between flavonoid and nonflavonoid polyphenols—we are starting with the latter group. Generally speaking, these are found in grapes and wines at low concentrations, with one exception: hydroxycinnamic acids are the major phenolics in white wine and are also found in red wine. Another group is the benzoic acids, such as gallic acid. The cinnamic and benzoic acids, also known as the "acid phenols," are small molecules and they are often present in grapes joined together with other chemicals. (This is known as a conjugated form. Examples would be as esters or glycosides.) These are easily extracted from the pulp of grapes during pressing and occur at levels of 50–250mg/l. Typical levels in finished wines would be 130mg/l. These nonflavonoids can oxidize in white wines and form a brown color and typically they taste quite bitter, especially when they are glycosides.

Other nonflavonoid phenolics found in wine include the hydrolysable tannins (such as gallic and ellagic acid), which come from oak barrels, the volatile phenols (such as 4-ethyl phenol, produced by *Brettanomyces*), and stilbenes (important in plant disease resistance, including one that is a bit of a celebrity, the phytoalexin resveratrol).

FLAVONOID COMPOUNDS—THE SKIN AND SEED PHENOLICS

These are the major phenolic compounds in grapes, and most of them are found in grape skins, but they also come from seeds and stems. This is where we bump into the two most famous types of the polyphenols—anthocyanins and tannins. The flavonoid phenolic are broken down into two groups: flavan-3-ols and flavonols.

FLAVAN-3-OLS

First, we have flavan-3-ol monomers. These are responsible for much of the bitterness in wine, and they may also have an astringent taste. This is one of the interesting things about polyphenols in general: they have both a flavor (bitterness is sensed by bitter receptors in the tongue) and a sensation (astringency is sensed by touch receptors and is not actually a taste). But more on this later. The major ones are (+)-catechin, (−)-epicatechin, and (−)-epicatechin-3-O-gallate. These are formed before *veraison* and change with ripening. They mostly come from seeds. Then we have polymers of flavan-3-ol subunits, which are generally referred to as condensed tannins or proanthocyanidins (or procyanidins). These are responsible for astringency and perhaps also bitterness in red wines and come from the hypodermal layers of the skin and also the parenchymal layer in the seed coat.

Tannins

This is where we consider tannins, which are pretty important, and really interesting. The name comes from the process of tanning leather. This is the process by which an animal skin is cleaned and made into a useful, durable material; one of the ways of doing this involved using plant tannins to adapt the proteins in the hide. Most of the subunits that make up tannins are either catechin or epicatechin and these can be modified in various ways. The length of this tannin chain—the polymer length—can vary from 2 or 3 subunits to over 30. The length of the chain is called the degree of polymerization (DP), so a 10 DP tannin has 10 flavan-3-ol subunits in it.

Skin tannins are usually much bigger than seed tannins, and they may contain some epigallocatechin subunits. Seed tannins are smaller and lack epigallocatechin subunits. They also have a higher proportion of epicatechin gallate, which is rarely found in the skin tannins. Skin tannins have a DP of up to 20, whereas the DP for seed tannins is up to 15.

Tannins in berries change during the ripening process. In the skin, there is little alteration in quantity of tannin from *verasion* to harvest, but the DP changes significantly. From green to red berries the average goes up from 7 to 11, and at harvest DP is around 20. But these are grape tannins, and recent science has shown that there is quite a journey from grape tannins to wine tannins.

First of all, the tannins have to be extracted from the grapes; secondly, they undergo many chemical rearrangements, and these changes keep occurring even after the wine is bottled. One of the key aspects of red winemaking is extracting the right sort of tannins, along with the other molecules that then can combine with tannins to produce stable color and a nice mouthfeel. The same is true for white wines where there is a degree of maceration, such as orange wines (see also page 147). One of the findings from a big, recent tannin study was that after fermentation the suspended grape pulp material can bind and remove tannin. This might explain why extended maceration of red wines post-ferment can change the way the wine feels in the mouth. In addition, they showed that different strains of yeast can vary tannin extraction during fermentation by as much as 50 percent. Higher alcohol may also increase tannin and anthocyanin (more on this later, see page 143) extraction. And it is quite common to fine red wines before bottling them. It used to be thought that the addition of proteins such as egg white or gelatin to a wine causes them to bind them to tannins and remove them from solution, thus improving mouthfeel by reducing astringency. But this process does not actually remove much tannin. What happens is that the proteins do bind to tannins, but these largely remain in the wine, forming colloidal or soluble complexes that likely have reduced astringency.

Tannins are largely responsible for the mouthfeel of red wines (and whites where there has been extended skin maceration). We sense them by a mixture of taste and also touch. Tannins can have a bitter taste, especially when they are smaller (with a low DP). But the main way we sense them is by touch: they are astringent. They bind to proteins in our saliva and then the tannin–protein complexes precipitate, giving a drying sensation in the mouth.

The key role of plant tannins is as defensive compounds. They are there to neutralize the nutritional benefit of eating the plant, by binding to digestive enzymes (which are proteins) of anyone who might want to feed on them. And this is one of the reasons that we have proteins in our saliva: to protect us from the harmful effects of tannins by binding to them and precipitating them before they reach the gut and begin deactivating our stomach enzymes. This makes plants more edible than they otherwise would be, neutralizing one of their defenses. The unpleasant taste of unripe fruits is in part due to high tannin concentrations, with the plant using this as a way of keeping the fruit from being consumed before the seeds are ready for dispersal, along with color changes and high acid/low sugar.

One of the protein types found in our saliva is mucin, which is involved in forming a lubricated, slippery protective layer over the internal surface of the mouth. Tannins remove this lubrication, causing a sense of dryness and puckering. This is what we describe as "astringent." It is one reason why tasting lots of red wines in quick succession can be an unpleasant experience. And if you have ever looked in a spit bucket when people have been tasting and spitting reds, then you will see streams of red-colored globs where the saliva and wine tannins (many of which are complexed with pigments) have combined in the most unattractive of ways.

Related to astringency is the taste of bitterness. The majority of tannins are chiefly sensed as astringent, but they can also be tasted as "bitter" when they are small enough to interact with bitter receptors on the tongue. Tannins seem to reach their most bitter taste at a DP of 4, and then decrease in bitterness and increase in astringency, with this astringency peaking at a DP of 7 (according to some studies, at least—others suggest it carries on increasing to DP 20), before growing steadily less astringent as they become larger. The astringent nature of tannins can be moderated by the presence of polysaccharides (sugars) or other wine components. It is also modified by the chemical adornments that tannins can grab, and there are many of these.

In wine, tannins are continually changing their length (DP) and adding things to their structure. So, structurally, wine tannins can be incredibly complicated, and researchers are still trying to correlate mouthfeel properties with structure. The conclusion so far is that interestingly, tannins are more astringent with lower pH (that is, wines with higher acidity taste more astringent, even with the same tannin content) and less astringent with increasing alcohol. But while low pH increases the perception of astringency, it does not alter the tannin structure. In contrast, the bitterness of tannins rises with alcohol level, and is unchanged by pH changes.

Anthocyanins

Anthocyanins are also flavan-3-ols and are the main pigments in wine, responsible for the color of red wines. They are found along with tannins in the hypodermal cells of the skin, except in teinturier (red-fleshed) grape varieties, where they are also found in the pulp.

Over 600 anthocyanins have been identified in nature, and these are formed from six different basic anthocyanidin structures, called aglycones. These six are cyanidin (Cy), pelargonidin (Pg), delphinidin (Dp), petunidin (Pt), malvidin (Mv), and peonidin (Pn). These all differ in color slightly, with some more red and some more purple. And so, anthocyanins are the combination of an anthocyanidin connected to a sugar (glycone) molecule, making the proper name for one grape anthocyanin malvidin-3-glucoside. Anthocyanins also differ in color according to the pH, with a more red color at low pH and a more blue color at high pH: this can be seen when you rinse out a wine glass with a tiny bit of red wine

in it with tap water: the rise in pH causes the trace of wine to go from red to blue-black.

These basic aglycones are not very stable but may be modified by chemical processes called glycosylation and acylation, which improve their stability. The acylation can take place with the addition of acetic, p-coumaric and caffeic acids. With these sorts of modifications, there can be up to 20 different anthocyanins in red grapes depending on the variety. It is interesting to note that Pinot Noir lacks acylated anthocyanins, which explains why Pinots are usually paler in color than most other red wines, and in contrast Cabernet Sauvignon contains an almost full suite of 18 of the 20 anthocyanins.

The anthocyanins can exist in several equilibrium forms in wine. One of these forms is called the flavylium ion form and this is very important because it is red colored. Another form is the quinoidal base, which has a blue color. Only a small proportion of anthocyanins are in these colored forms in wine. It follows that if more of the anthocyanins are in the flavylium or quinoidal form, the wine will have more color. When sulfur dioxide is added to wine it has a bleaching effect because it binds some of the anthocyanin as a colorless bisulfate adduct.

Anthocyanins are unstable in wine and are not that important for the long-term color of red wines. Recent studies have shown that after four years in bottle, there are no colored anthocyanins left in red wine. So, what is it that is causing the color in the wine, then? Well, in addition to the anthocyanins there are two major fermentation and aging-derived color groups. The first of these is the pigmented polymers. These are formed by the chemical linkage between tannins and anthocyanins. This is a covalent (strong) linkage and is very important in forming stable color in wines. The evidence suggests that most of the pigmented-polymer formation occurs during fermentation, but according to some reports by the end of fermentation around 25 percent of the anthocyanins are thought to be complexed with tannins, and in barrel-aged reds this figure can rise to 40 percent within a year. Barrels help because they provide a bit of exposure to oxygen, which helps in forming these complexes, as well as supplying some extra tannins. Acetaldehyde, the product the oxidation of alcohol, helps to form these bonds. During fermentation, the formation of tannin–anthocyanin complexes helps retain more of both tannins and anthocyanins in the wine, as it makes them more soluble and stops them from dropping out as a deposit. So, having more anthocyanin present means that more tannin will be retained in the wine, which suggests that teinturiers might be useful as a blender in a co-ferment, if more structure and color is desired. These polymeric pigments also help with color intensity: around two-thirds of the anthocyanins are in a colored form when they are complexed with tannins, as opposed to around ten percent of the free anthocyanins that are in a colored form.

Then there is another group called the anthocyanin-derived pigments, which arise from reactions between anthocyanins and other phenolics and aldehydes. Called the pyroanthocyanins, this is a large, complicated class of pigments, and is an area of intense current research, with new members being added all the time. These pyroanthocyanins include the vitisins, portisins, and oxovitisins. They are stable and are resistant to sulfur dioxide bleaching. Most of them have a yellow/orange color with the exception of portisins, which are blue.

The phenomenon of co-pigmentation needs a mention. The colored anthocyanins (red flavylium or blue quinoidal base) are planar structures and these can react with other planar structures (in this case these will be referred to as co-pigments) such that they form a molecular stack that excludes water. This protects the anthocyanins from hydration, increases the color intensity, and shifts the color toward purple. The co-pigments are usually other phenolic compounds, in particular

the flavonols, which we will discuss below. This is one of the reasons that red grapes are sometimes co-fermented with a small proportion of white grapes: the white-grape skins provide co-pigments, so even though you would expect that including white grapes would reduce the color of the wine, they do not. In fact, the opposite occurs.

The classic combination is a small proportion of Viognier together with Syrah/Shiraz. The flavonol concentration is also increased by UV light exposure, so you might expect grapes that have experienced higher sunlight exposure to show a deeper color through increased co-pigmentation effects. While co-pigmentation can increase the color intensity of young wines, this effect does not persist with bottle aging. It is lost over time because the anthocyanins have been converted to more stable wine pigments.

FLAVONOLS

The final group of phenolic compounds we will look at is the flavonols. They are found in the skins of both red and white grapes and act as sunscreens against UV-A and UV-B light wavelengths.

Flavonol levels increase in response to enhanced UV exposure. They have a yellow color that can contribute to the color of white wines, but which is masked in red wines. The most important of the flavonols is quercitin, but kaempferol, myricetin, laricitrin, isorhamnetin and syringetin are also found. White grapes lack myricetin, laricitrin, and syringetin. They have high antioxidant capability, but perhaps their most important role is as acting as co-pigments with anthocyanins to increase the color of new red wines.

PHENOLICS AND OXIDATION

Phenolic compounds have an important role to play in oxidation. In order for wine to oxidize, phenolic compounds and also transition metal ions (such as copper and iron) are needed. The most important phenolics for oxidation are the o-diphenols (ortho-diphenolic compounds,

ABOVE Must hyperoxidation. Allowing the must contact with air oxidizes the phenolics, turning it an alarming-looking brown color. But this takes the phenolics out, making the wine much less sensitive to oxidation later on.

also known as catechols). These include gallic acid, caffeic acid, caftaric acid, epicatechin and catechin, and all flavan-3-ols. Remember the distinction between pulp phenolics and skin/seed phenolics? The hydroxycinnamic acids (hydroxycinnamates) are the most common of the pulp phenolics, and so these are frequently the phenolics most commonly found in white wine. Caftaric acid is the most widely found hydroxycinnamate found in grapes and it consists of caffeic acid bound to tartaric acid, and in wine this is hydrolyzed by yeasts during fermentation to form caffeic acid, which then becomes the most common hydroxycinnamate in wine. Catechin and epicatechin are the basic units of grape and wine tannins found in the skins and stems. If you press more heavily you get more catechin and epicatechin in the wine, and if there is skin contact there is more in the wine, too. Their concentration correlates with browning susceptibility in white wine. Most of the literature suggests that the skin/stem phenolics are responsible for browning potential of white wines, while some studies suggest that the level of o-diphenols (including pulp and skin/stem phenolics) in white wines is correlated to their ability to brown. Overall, the evidence seems to point the finger at the skin/stem phenolics as being those most responsible.

In the pressing of white wines, two approaches are often taken. The first is to protect the must from oxygen, but to only press lightly, which results in low levels of phenolics in the wine. The second is to press in the presence of oxygen, where any phenolics in the juice will oxidize and fall out. The juice turns a horrible brown color but then when it has fermented, because of the low levels of phenolics, it is less sensitive to oxidation. Wines protected too much from oxygen early in their life but which have high-ish levels of phenolics can run the risk of oxidizing more easily once bottled.

11 Extraction and maceration

It was a balmy September evening in the upper reaches of Portugal's Douro Valley. I was with a small group of journalists, staying at Quinta de Vargellas, the well-known Port quinta owned by Taylor Fladgate. We finished dinner, slightly merry after a few glasses of Port, and headed down to the winery. As we approached, we could hear the sound of music. It sounded like a party. To an outsider, what we were about to witness might seem quite bizarre, but anyone with a little of knowledge of the Douro would realize this is a bit of living history. Inside there were two stone *lagares*, each with maybe a score of people dancing to the music knee deep in a viscous pool of a dark-red/black liquid. They were foot-treading grapes.

The traditional vintage practice in the Douro Valley is to invite the inhabitants of a village to come stay at your quinta. Then, during the day

they pick the grapes and bring them back to the winery. These are placed in the shallow granite troughs called *lagares*. In the evening, it is time to tread them. The human foot is particularly good for treading grapes, because it exerts enough pressure to crush the skins and pulp, extracting all the good things, while leaving the stems and seeds—which have the potential to add bitterness and greenness—intact.

The challenge for Port-making is to extract everything from the skins in a very short window of two to three days. For making Port wine, the fermentation is stopped while there is still lots of sugar left in the wine by adding colorless, high-strength brandy. This process is known as fortification. Because of this, the skins and the juice are only together for a few days, so it is necessary to extract as much from them as quickly as possible. Each evening there are two hours of regimented treading called the *corte* or cut, and then there is free treading (usually with dancing) called *liberdade*: the music starts and it's a party atmosphere.

The crushed grapes float in a viscous mass of intensely colored liquid, moving slowly in waves with the foot treading. We take our shoes off, hose down our legs and feet, and get in.

It's a wonderful feeling treading the grapes, which by this stage have become mushy and fragmented. Fermentation on the skins is brief, because when the sugar level has more-or-less halved from 240g/l to around 120g/l, the Port is fortified with spirits. Thus, Port is one of the extremes of extraction: a lot, in a short period, but as gently as possible.

Above Foot treading Port in a *lagar* at Quinta de Vargellas in Portugal's Douro wine region. The human foot is a great way of crushing and then extracting the color, tannins, and other good components from the skins of the grapes, without damaging the seeds that could release bitterness.

It is becoming increasingly difficult for Port producers to find people to do this work. The search for alternative ways of getting this rapid but gentle extraction has led to the development of robotic *lagares*. These are quite impressive to watch in action. They have silicon bungs on the bottom and press down with similar force to a human foot.

EXTRACTION IS AN ESSENTIAL PART OF WINEMAKING

At least for red wines, rosés, and some white wines. Extraction refers to taking material out of the skins of grapes. Standard protocols for white winemaking tend to minimize extraction from the skins, achieved by pressing the grapes soon after harvesting. They might be whole-bunch pressed, where entire clusters are put into the press, or they might be destemmed with or without crushing first. Sometimes there is deliberate skin contact for a few hours, in order to get some flavor components out of the skins. But in any white winemaking, there is inevitably some extraction, and the degree to which this occurs is important for both style and quality.

Recent years have seen the emergence of a new class of white wine, termed orange or amber wines where fermentation and sometimes aging also take place on the skins, as with red wines. Clearly, extraction is a big issue here. And extraction is a big deal for making rosé wines, because some, but usually limited, color is wanted from the skins of red grapes to produce a wine with the desired color. Here, though, I will begin with red winemaking, and the role of extraction, and how it relates to color.

For making red wines, there are lots of choices about how, and how much to extract from the skins. To begin this journey, let's consider the composition of a grape berry.

UNDERSTANDING THE GRAPE BERRY

We will begin on the outside, with the skin. Also known as the exocarp, this consists of two layers.

First, there is the epidermis, a single layer of cells covered in a waxy cuticle. This cuticle consists of overlapping plates of wax that are one-third soft wax and two-thirds hard wax. The epidermis of grape berries is unusual in that it has very few stomata (the gas exchange pores common on other plant extremities). Below this layer is the hypodermis, which is a few layers thick. Together, these form what we call the grape skin. The skin contains around 35 percent of the organic acids of the berry, 25 percent of the sugars, and 20–30 percent of the phenolics. Significantly, it contains over 80 percent of the aroma compounds, and also it is where the pigments—mainly anthocyanins—accumulate in red grapes after *veraison* (where the skins change color and begin to soften). Below the skin, we have the pulp of the grape, also known as the mesocarp. This has bigger cells and is not colored (except in red-fleshed teinturier grapes). The pulp has 50 percent of the sugars, 35 percent of the organic acids, less than 20 percent of the aromatic compounds and 10–20 percent of the phenolics (these are the phenolics found in white juice, therefore). Then we have the endocarp and seeds in the center of the berry, which have 25 percent of the sugars, 30 percent of the organic acids, and 60 percent of the phenolics.

For making white wines, the first pressing will release the juice from the mesocarp. Press harder, and you'll get juice from the endocarp, and some extraction from the skins. Press harder, and it will mainly be more extraction from the skins. Thus, harder pressing gives more phenolics. It also gives more potassium, because the skins have the highest concentrations of potassium.

For red winemaking, a lot of the desirable components are present in the skins, and so extraction aims to get them out. Ideally, you want to avoid the 60 percent of the phenolics present in the seeds/endocarp, because these are of a different composition to skin tannins and are thought to be more bitter. Some will be extracted, of course, but it is good to keep this to

Above Working the ferment at Quinta do Panascal in the Douro, Portugal. For Port wine, extraction has to be rapid because the wine only spends a few days in contact with the skins before being pressed. After fermentation has progressed just a little, spirit is added.

Above Pumping over a fermentation at the Pintia winery, Toro, in Spain. The wine is pumped up from the bottom of the tank to the top. The idea is to keep the cap of skins wet to stop it from going acetic, while also gently extracting the good things from the skins.

a minimum. This is a subject we'll return to when we discuss extended maceration (see page 150).

There are various factors influencing what is extracted from the skins. The temperature (higher temperatures extract more), the presence of alcohol, the maturity level of the grapes, the physical handling (punch down, pump over, heading down, rack and return, rotofermenter), the shape of the fermentation vessel (and particularly its diameter relative to its height), and the presence or absence of enzymes.Traditional red winemaking is thought to extract about 20–40 percent of grape phenolics into the wine.

HOT EXTRACTION: THERMOVINIFICATION AND FRIENDS

The idea of heating grapes to make it easier to extract from the skins is an old one. Dr. Simon Nordestgaard of the Australian Wine Research Institute has charted the history of this technique and finds references to this from the 18th century. Nordestgaard notes that in the 19th century Prunaire talks of pre-fermentation heating of grapes in Burgundy to get better color. In 1906, Bioletti published a report on a new method

for making red wines involving a device that heats up must using copper pipes and steam, with the hot must then being used to extract material from the skins. In the 1920s, Ferré came up with a new method in Burgundy that heated entire grapes, but this was not widely adopted.

Things changed in the 1960s in France, when a series of poor vintages led to increased interest in heat-treating grapes. Here, the prime motivation was not necessarily extraction, but inactivating the laccase enzymes in the skins of grapes that had been infected by *Botrytis cinarea*. These are members of a group of enzymes called polyphenol oxidases that cause enzymic oxidation of grape musts after crushing or pressing—a process much more rapid than chemical oxidation.

The principle behind hot extraction is that grapes are heated before fermentation in order to facilitate extraction. The grapes are typically heated at 50–90°C (122–194°F) and held for a period. Then you press the grapes—either immediately or after a short fermentation on skins—and do a liquid-phase fermentation off the skins, as you might with a white wine. The idea is that this rapid extraction allows you to take out what you want: anthocyanins and aroma compounds and precursors, and just a bit of tannin. Doing fermentation in the liquid phase means it is easy to control fermentation temperature. Another benefit is that if the grapes are not in perfect condition because of disease, the high temperatures will neutralize the oxidase enzymes and because the skins are not macerated for long it will minimize the pick-up of off-odors from damaged grapes. One potential difficulty with short high-temperature maceration is that anthocyanins are extracted, but not many tannins. The tannins are needed to stabilize the anthocyanins by forming pigmented polymers.

In France, the term thermovinification is reserved for short (one hour or so) high-temperature macerations, while MPC (*macération préfermentaire à chaud*) describes pre-fermentation at 70°C (158°F) for a longer time (typically 12 hours) followed by either a solid- or liquid-phase fermentation. In Germany, there is a technique called KZHE, which is a short high-temperature treatment (2 minutes at 85°C/185°F), followed by 6–10 hours at 45°C (113°F), then a liquid-phase fermentation.

Flash détente (flash release) is a heat-extraction technique patented by INRA (Institut National de la Recherche Agronomique) in 1993. It involves heating the grapes to 85°C (185°F) for a short time by direct injection of steam, and then using a vacuum to cause the water inside the cells to boil making everything much more extractable. There are two ways of doing this: cooling down to 30°C/86°F (full), or cooling down to 50°C/122°F (half). This can reduce green characters and is good for grapes that are not fully ripe. It also helps to maintain the varietal profile of the wine, which can be lost with some of the other thermovinification techniques: a common problem cited with these is standardized aromas and a banana-yogurt character. In France in 2008, 500 million liters

(110 gallons) of wine were hot extracted. I am not aware of any more recent data points.

THERMOVINIFICATION IN CHAMPAGNE

One surprising use of thermovinification is by Champagne house Moët & Chandon. They use it to make the red wine component of their Imperial Rosé, which they made for the first time in 1996. They now have four rosés in the range, and these represent 20 percent of their production. "In twenty years we have gone from two to twenty percent," says chief winemaker Benoît Gouez, "whereas in the region as a whole it has gone from two to ten percent." To make this change they need to make a lot of red wine, and they have invested in two red wineries, one in Épernay for making Pinot Meunier from the Marne, and the other in Val du Clos in the Aube to make Pinot Noir.

Thermovinification was introduced at Moët & Chandon back in the 1970s, imported from the south of France. The purpose was to have a way of making red wines even in bad years. "This was the idea in the 1970s, making red wines from grapes not perfectly clean," explains Gouez. At that stage, it was common to have lots of rot-affected grapes coming into the winery. Now though, the grapes are in much better condition, so there is a second reason they have brought thermovinification back. This is because it also explodes the cells of the skins, which releases anthocyanins and aromatic precursors, so they only extract for two hours with agitation and then press. Then they settle and ferment in the liquid phase at 18°C (64°F) like a white, rather than fermentations on skins, which are usually at 28–30°C (82–86°F). The reason for this is to have very little tannic extraction, with no seed tannins at all. They have also discovered that they can extract and then reveal a lot of thiol precursors. "These are mostly known in Sauvignon Blanc, but we have realized that in Champagne, in Pinot Noir, Pinot Meunier, and Chardonnay, that we have the thiol precursors," says Gouez.

"That character of rhubarb is an expression of 3MH and 3MHA." These polyfunctional thiols add aromatic interest but are relatively short-lived in wine. Thermovinification has become a signature of their NV Rosé. "We are basically the only ones who use this technology in Champagne. It is what we are looking for: we want to keep Rosé Imperial light and easy to drink without much structure. We keep the tannins and the structure for the Vintage." A large quantity of red wine is added to the rosé. For Vintage it is nine percent and for NV thirteen percent. But ten years ago, when the red wines were paler, they added twenty-five percent.

PRE- AND POST-FERMENT MACERATIONS

One technique in red winemaking—and now with the advent of orange or amber wines, also with some whites—is post-ferment maceration on skins. If the extraction that goes on during Port winemaking—a quick two or three days mostly in the aqueous phase—is at one extreme, then post-ferment maceration with the skins in contact with juice and wine is at the other. This can take place for as long as a year in some cases, but more typically will last 30–90 days.

In Georgia, in Eurasia, typical *qvevri* winemaking will involve filling these large underground clay vessels with intact grape bunches, sealing them up and then leaving them. Typically for reds, this will be for one to three months, but for whites it is actually longer: six months is common. The shape of the *qvevri* lets the seeds accumulate at the bottom with limited contact with the juice, and this is thought to reduce extraction of seed tannins.

I spoke to Yalumba winemaker Kevin Glastonbury who has been doing post-ferment maceration on Grenache. About one-third of the Yalumba Bush Vine Grenache will spend a reasonable length of time on skins post-ferment. "A couple of the batches will stay on skins and the rest will go straight to barrel," says Glastonbury. "In this vintage [2018] a third

of the wine was on skins for sixty-one days. The reason we do this is stylistic: we are looking for a tannin integration with the phenolics, to give a rich savoriness, but juicy still. The rest of the wine is matured in old barrels. We are looking for a red-fruited wine with mid-palate flesh."

So, what happens to the tannins in a post-ferment maceration? I asked Glastonbury. "I don't know chemically what is happening other than there is a lot of binding of many things going on. One of the intriguing things about tasting a Grenache as it evolves post-ferment, is that in some weeks you can find the wine tasting quite thin on the palate: it is almost as though it has been watered down. Then, a couple of weeks later, you can taste it again, and find that it is very bitter, with a green peppercorn bitterness. And the next week it is balanced, even, and fleshy. It goes in waves like this. How and why it should do this, I am sure there are documents on it. But I live in the world and reality of tasting wine. I am there to taste the wine and when I think it is ready to come off, when it is the right spectrum of balance and flesh, that is when we will take it out. Certainly, there is binding of phenolics, anthocyanins, and tannin compounds."

THE EFFECTS ON TANNIN EXTRACTION

Extended maceration increases the extraction of seed tannins. These are found in a layer of cells underneath the cuticle of the seed, and are formed of catechin, epicatechin, and epicatechin-3-O-gallate, mainly present as monomers (single units), but also as polymers. Studies have shown that seed tannins are shorter (with a lower degree of polymerization) than skin tannins, which means that they tend to be more bitter and astringent. But the extraction of these seed tannins follows an unpredictable course in maceration.

Extraction of skin tannins occurs early on in maceration. Again, there is not a simple relationship between maceration time and tannin and anthocyanin levels in the wine. Anthocyanins are lost during maceration, but they increase the solubility and thus retention of tannins in the wine by the formation of pigmented polymers. These make the color more stable and also change the mouthfeel of the wine.

In a 2013 paper, Federico Casassa and colleagues in California looked at the effects of extended maceration and different alcohol levels on winemaking. Two lengths of maceration (10 and 30 days) and two alcohol levels (12 and 13%) were used.

They followed the wines for a year after bottling, looking at the chemistry and sensory changes. They found that aside from the alcohol level, it was the maceration period that defined the chemical change. The tannin in extended maceration was mainly from seeds, while in the control wines it came from seeds and skin in equal measure. One result that was interesting was that tannins recovered in the skins after maceration had a significantly lower degree of polymerization than those in the skins of the fruit prior to maceration. The evidence suggests that the seed tannins might be binding to the skin in the pomace. This fining of seed tannins would be one explanation why extended-maceration wines tend not always to taste more tannic than shorter macerated wines: the skins are binding some of the seed tannins, which are therefore removed from the wine at the pressing stage. In the sensory analysis in this study, the extended maceration wines were not perceived as being more bitter, but they were rated 30 percent more astringent.

Pre-ferment macerations at low temperature are also common. The idea here is to begin extracting the good stuff in the skins in an aqueous environment, before any alcohol—which acts as a solvent—is present. Anthocyanins are more soluble in must than tannins, so they are typically extracted more in this phase.

Venetia Joscelyne from the University of Tasmania did her Ph.D thesis on pre- and post-ferment maceration techniques. She carried out a survey in 2005 asking 700 winemakers in Australia about their use of extended maceration and pre-ferment maceration. She got 103 replies: 57 percent did pre-ferment maceration, and 61 percent post-ferment. For pre-ferment, the duration was under 3 days for 42 percent. Typically, they would add 50ppm sulfur dioxide at the crusher. For post-ferment, the majority did one to four weeks, but some went as long as six to seven weeks. The winemakers were under the impression that it reduced the bitterness and astringency in the wine, and they didn't use it as much as they would like because of harvest logistics—it ties up a fermenter.

In the next chapter, I will look at the topical and very interesting subject of whole-bunch fermentations, including another form of maceration: carbonic.

12 Whole-cluster and carbonic maceration

The world of fine wine is big and diverse, and it is dangerous to make generalizations. But if we take a wide-angle look, winemaking trends seem to move in cycles. It seems we are currently in a phase where elegance and complexity are being pursued by winegrowers at the expense of power and strength. Visit wine regions worldwide and you'll find very few young winemakers aiming to produce bigger wines—certainly not at the high end. They tend to prize elegance, freshness, and definition above all else. The monster 100-point wines of the 1990s and 2000s are increasingly looking like yesterday's wines.

It is perhaps for this reason that there is increasing interest in winemaking techniques that foster this elegance and complexity. For example, there's a marked shift away from small new oak as the primary vessel of *élevage*, with renewed interest in concrete and larger, more neutral oak. Wild ferments used to be a novelty; now they almost seem normal. While many are suspicious of the natural-wine movement, even those people outside of it have begun to work more naturally in the cellar because they believe this is likely to result in wines that express their sense of place better. And there is increasing discussion of and experimentation with the topic of this article, that is, the use of stems in making red wines.

WHOLE-BUNCH OR WHOLE-CLUSTER FERMENTATION

There is, of course, nothing new about making wine with the stems included, which is usually referred to as whole-bunch or whole-cluster fermentation. Through history, wines would have been made from intact bunches that were then either pressed immediately to yield juice for white-wine fermentation or macerated during the fermentation process for red-wine production. The only way to remove the stems prior to red-wine fermentation would have been manually picking off each berry or using some sort of screen with holes in it that is placed over a fermenter: the bunches are then pushed across this and the berries fall out into the vessel below. This is time-consuming and therefore very expensive, although there is one famous Bordeaux estate that practices it for its first wine, Château Pape-Clément, and in Burgundy Domaine de la Vougerie does this on a smaller scale for their Musigny Grand Cru. In Chile, Lapostolle's Clos Apalta is also made from grapes that are destemmed by hand. But the development of the crusher destemmer allowed winegrowers a quick, economical way of separating out the stems from the berries, and the vast majority of red wines are now made from grapes that are first destemmed, either by such a machine or, increasingly, in the vineyard by the machine harvester.

Some anatomy. When a bunch of grapes is picked, it consists of the grapes, plus some other material that holds the cluster together. The main axis of the cluster is the rachis, and the pedicel attaches the berries to this. The part that attaches the cluster to the vine, and which is cut through to release the bunch at picking, is called the peduncle. Together, this material, which we are referring to with the broad term "stem," consists of about two–five percent of the weight of the cluster. Depending on the region and that year's climate conditions, stems can vary a lot in their appearance. This is because they start out as green photosynthetic material, and then undergo a process of lignification, which is the transition from green fleshy plant to woody plant, achieved through the deposition of lignin in the spaces in the cell walls between the cellulose fibers. So, having stems in the fermentation can mean very different things in terms of wine outcome, depending on the degree of lignification of these stems.

WHO DOES WHOLE-BUNCH?

Burgundy is the region most associated with whole-cluster fermentation and by association Pinot Noir is the grape variety most closely linked with this technique. In part, this could be because Pinot Noir as a variety lacks acylated anthocyanins, which are a form of pigment. This explains why Pinot Noir is usually lighter in color than other red wines, but in addition, anthocyanins also have important interactions with tannins in wine, and form pigmented polymers, which are important in wine structure and color. Some of the wood tannins leached from the stems could be making up for this shortfall in Pinot Noir. But traditionalists in the Northern Rhône with Syrah have also used it. Increasingly, New World producers working with

Pinot Noir have been exploring the use of whole clusters, and it is also beginning to catch on with Syrah producers who are looking for elegance.

The Burgundian domaines most famously associated with whole-bunch are Romanée-Conti, Dujac, and Leroy. "Clearly, in Burgundy at the moment there is a tendency to move toward stems," says British wine writer and Burgundy expert Jasper Morris. "I can see two main reasons for this," he says. "One is that Henri Jayer, who hated stems, is dead. And the other is that with global warming, the stems are more often riper than they used to be." Jayer, a tremendously high-profile grower, influenced many to move away from stems and, until recently, this was the direction being taken across the region. And the popularity of destemming was linked with a corresponding reduction in greenness and rusticity in many red Burgundies, so there was a good reason for doing it. In a sense, in the past people used stems by default, and the results weren't always good. Now the choice to use stems is an active one, so the people doing it are doing a better job with it.

Jeremy Seysses at Domaine Dujac uses between 65- and 100-percent whole-cluster fermentations, depending on the cuvée. "We have the feeling that we get greater complexity and silkier tannins with whole-cluster fermentation," he shares. "In

Left A whole-bunch fermentation photographed halfway through. This is at the Gabrielskloof winery in Bot River, South Africa. The wine in question is the Crystallum Whole Bunch Pinot Noir 2018, which turned out very well.

high-acid vintages, it helps round things out, and in high-ripeness vintages, it brings a freshness to the wines." For Seysses, the decision about whether or not to destem depends on a number of factors. "Some terroirs don't seem to do so well with whole-cluster. The whole-cluster character rapidly becomes dominant and can appear 'gimmicky,' it doesn't mesh well with the wine, and can give the illusion of complexity, but it feels superficial," he explains. "Of our holdings, I like destemming a little more for the Gevrey vineyards than the others." He also tends to destem more frequently the grapes from younger vineyards with bigger clusters, and in vintages with rapid end of season ripening, where the ripening may be a little more uneven.

Mike Symons, winemaker at Stonier in Australia's Mornington Peninsula region also finds that terroir is the biggest determinant of whether or not he uses stems in his ferments. "We have a couple of vineyards, where we like the stems, that are north-facing and produce nice ripe stems," he explains. "We pretty much know the vineyards where we like the stems. One of them is the Windmill vineyard, and another is the vineyard near the winery. There are some vineyards where we don't include the stems, such as the Lyncroft vineyard, which is very cool, or the KBS vineyard Pinot. They would just be awful if we included the stems."

"I normally find a strong correlation between the better sites and the amount of stem/whole-bunch I am able to use," says Mark Haisma, an Australian working as a micro-négociant (someone who specializes in limited-production wines) in Burgundy and Cornas. "The stems from the best sites are generally cleaner and richer in character."

To decide whether or not to use stems, Mike Symons eats them. "If they taste like broccoli we don't use them." His experience from regular workshops with Victorian Pinot Noir producers is that winemakers are increasingly talking about using stems, but he thinks that

you need the right vineyard. "Some people get on bandwagons and they include stems where they shouldn't. It is something you have to be careful of," says Symons. "With a blend where we include stems, we will do it over three days or more to make sure we get it right."

Another well-known Victorian winemaker, Tom Carson of Yabby Lake, admits that he likes to play around a bit with whole bunches in his Pinot Noir ferments. "I am still experimenting, and I'm reluctant to go in too hard. When it's good, whole-bunch fermentation gives fragrance and perfume, and adds a bit of strength and firmness to the tannins. But when it's not good it can dull the fruit, adding mulch and compost character," says Carson. "We want to highlight the fragrance of the Pinot. We don't want complexing elements that are not vineyard-derived." Carson did eight percent whole-bunch in 2009 and twenty percent in 2010, but then backed off a lot in 2011 because it was a wet year and the stalks were very green. "We are still learning what is the right amount."

Nick Mills, of Rippon, in New Zealand's Central Otago, uses some whole-bunches in the Pinot Noir ferments, but decisions are made based on the fruit. "We do some whole-bunches," says Nick, "but this is all done on the sorting table." He adds that, "the sorting table isn't about taking stuff off, but it's for me to taste pips and skins, and figure out what raw material we have. If we can chew the stems through then we'll put them in. I'd put in a hundred percent whole-clusters if we could. It's a better ferment." Overall, Rippon Pinot Noir has 25–40 percent whole-clusters. "The vineyard is incredibly parcelated," says Nick, "with all these small micro-ferments. If we get something really good, then we'll put the whole lot in and do a hundred percent stems, but if grapes come in that I don't like the taste of we'll use no stems."

It seems that lots of growers, like Mills, will use as many stems in the ferment as they can, with the limiting factor being the suitability of the stems. "When you are choosing whether or not to use stems, some people do a positive selection on the sorting table (*tri au positif*) rather than a negative one," says Jasper Morris. "So, when they come across bunches with lovely bronze stems, they use them."

Initially, Eben Sadie of South Africa's Swartland region didn't use any stems in making his celebrated Columella wine, but he decided to change this with 35 percent of stems included in the 2009 vintage. The goal was to get more freshness in a warm-climate wine. More recently, for the reds in his old-vine series of wines, he took whole-bunch to 100 percent before dialing it back to 85. "The only reason we destem 15 percent is to get some juice into the tank to have some layer of protection, to get the ferment going," says Sadie. "We did a hundred percent whole-bunch from 2011 14, but we started running into bacterial problems. We had a bit of *Brettanomyces* [brett], too. I don't like brett at all. I don't want any of it: it's not part of the terroir for me, although it comes on the grapes. Brett is a weird thing. I kind of like brett, but in other peoples' wine. It is like Stelvin: I like it on other peoples' wine."

THE EFFECT OF INCLUDING STEMS IN FERMENTATION

Stems have a number of effects on fermentations, but this is where the story becomes complex and somewhat unclear. There are many different ways of using stems in the fermenter, and the stems themselves can be quite different in terms of how green or lignified they are. "There is an immense difference in flavor profile from all the people who do use stems," says Jasper Morris, referring specifically to Burgundy. "You also have to look at the techniques involved. Here it gets very complicated." Morris adds that, "the stems in the fermenting vat will have perhaps a chemical impact, and certainly a physical impact."

"In small vats, like those used in Burgundy, stems are useful because they drain the juice in a more homogeneous way and keep the temperature of fermentation one or two degrees lower," says French wine commentator Michel Bettane. Jeremy Seysses agrees: "The cap is far more aerated, meaning that it doesn't get quite as crazy hot as it would without any rachis (the main part of the bunch stem) in there, letting some heat escape. It also drains much better when you punch down or pump over as you get no clumps." Nick Mills of Rippon in Central Otago adds that the presence of stems allows the yeasts to move around more easily, and the pressing is better. And Rhône winemaker Eric Texier claims that in whole-bunch fermentation, the conversion factor of sugar to alcohol is slightly different, resulting in wines with lower alcohol.

In addition to these benefits, Bettane adds that stems in the fermentation can also help diminish the negative influence of any fungal infection on the grapes. "In 1983, for instance, curiously the whole-bunch-made Burgundies were less flawed by rotten berries than destalked ones." But, he points out, if large tanks are being used, it is impossible to keep the stems, because they make the cap too resistant to mechanical pressure. Seysses also says that whole-bunch ferments are harder to punch down. "You have to do it by foot or by piston, you can't do it by arm. All these things change your extraction profile."

Another physical effect of stems in the ferment is a loss of color. "The stems also absorb color, leaching the color of the wine," explains Eben Sadie. "These days everyone wants to make more powerful, impressive wines, so whole-bunch is an unfashionable move because your wine looks weaker. For many people, color is an important property of the wine." But Sadie doesn't see this as a big problem. "I'll lose some color to gain freshness and purity. The wine has more vibrancy and life in it. Where we work in South Africa, the biggest flaw is our wines are often too ripe. It's good to get our wines fresher and more vibrant."

In addition, stems raise the pH of the wine slightly (making it less acidic, usually a bad thing), because of the presence of potassium released by the stems. The potassium then combines with tartaric acid and precipitates the acid out of the wine. However, Seysses points out that there is less potassium in the stems these days and so this not so much of a problem. (The potassium got there in the first place because a lot of Burgundy growers used too much fertilizer with high levels of it in the 1950s and 60s.) Some of this pH rise might also have to do with the carbonic maceration element that is a part of whole-bunch fermentations. This is the enzymic activity that occurs within the intact berries, which includes some transformation of the malic acid.

There are many variables involved in how the actual stems are added to the vat. "One question is, if you don't use all stems—and there is probably quite a lot to be said for using just some stems—when do you put them in?" asks Jasper Morris. "Do you put them in first, as a sort of base to the cuvée, and then you put your destemmed grapes on top? Or do you put them in last, so the stems slowly float down through the juice? Or do you do some sort of lasagne-like layering between stems and nonstems, which I have heard some people do?"

"With many cuvées and the destemmed fraction being so small, I inevitably co-ferment," explains Jeremy Seysses. "The practicalities of harvest don't always allow it, but I usually like putting the destemmed fruit at the bottom and the whole-cluster on top, so that it really stays whole. And as it can take a few days for the ferment to get going, I don't want my healthy whole-clusters to be covered with juice as they sit waiting for the yeasts to get going."

Another advocate of putting stems on top of destemmed fruit is Sebastien Jacquey of Megalomaniac winery in Canada's Niagara region. "I have been consistently using five–ten percent of stems in Pinot Noir, but I go to nothing when I have powdery on the stems," says Jacquey. "I usually do it on the last layers as I fill my tanks, and then put some crushed grapes on top. This is because

of Ontario: the Pinot Noir here is so sensitive, and by doing it like this I have the option to crush the whole-bunches if it starts going wrong." Jacquey also uses quite a lot of whole berries, handling them carefully to keep them intact. "It's mostly what I do," he says. "I pick early in the morning and destem gently. I go to a stainless-steel bin and then fill my tank using a forklift. I don't go through any pump. I try to have this intracellular pre-ferment: I like what the *Klockera* [wild yeasts] are doing early on, giving this forest-fruit character. I also like picking at different times, with some tanks with fresher, crunchy fruit, a little less opulent. And I have a tank that's a later pick, and then one a week and a half after the first."

Tim Kirk of Clonakilla, from Australia's Canberra District, uses some whole-bunch to make his Shiraz-Viognier, but as with many who do partial whole-bunch, they go in first. Whole-bunches are put into two-ton fermenters, only part full. Some Viognier is typically crushed and destemmed and put on this, and then some Shiraz is destemmed and put on top. Kirk reports that some of the grapes in the whole-cluster portion stay attached to the rachis and don't burst: he estimates around 20 percent of them. Instead, these berries begin fermentation from inside, as in carbonic maceration. If you take these whole berries out partway through fermentation, their pulp is colored red, so they are extracting color from the inside. They are also still a little sweet and, on pressing, these berries release sugar, which acts to prolong fermentation.

Blair Walter, of Felton Road in New Zealand's Central Otago, uses a little whole-bunch to add complexity to his wines, and he notes this delayed sugar release from intact berries. "We typically put in a quarter whole bunch and destem the rest of the bunches. And then when we punch down we don't go to the bottom of the tank. After twenty-eight days you can still pull out whole-bunches. They have fermented inside [the intact berries] and there is still some sweetness that is pulled out." He thinks this remaining sweetness is important because it keeps fermentation ticking along for a while. "Burgundians typically *chaptalize* in six to eight small additions," claims Walter. "This results in a slightly stressed fermentation producing more glycerol. This changes the texture and adds some fruit sweetness. It surprises me that more people don't use whole-bunches."

This partial carbonic-maceration character is likely to contribute significantly to the enhanced texture and aromatics often seen in wines made by whole-bunch fermentation. But Michel Bettane thinks that some of this benefit can also be derived from very careful destemming. "Don't forget that new destemmers are so precise and delicate that they allow winemakers to put 'caviar' destemmed berries in the vats with almost the same effect as whole-bunch fermentation," says Bettane. "The beginning of the fermentation takes place inside the berry, helping to preserve the best quality of fruit, delicacy of texture, and capacity to age, keeping the youth of fruit and avoiding barnyard undertones."

Mark Haisma is a winemaker with broad experience across different hemispheres. In his previous employ he was at Yarra Yering, in Australia's Yarra Valley, but he's now a micro-négociant in Burgundy, also making a wine in Cornas in the Northern Rhône. At Yarra Yering he developed an innovative approach to stem use, which he calls a "macerating basket." "The fruit would be completely destemmed, and I had some stainless-steel mesh cylinders made," he explains. "These would be stuffed with the stems. I could take them out when I felt I had what I wanted." I asked him about the results of using stems this way: "I find it adds a great spicy complexity to the wine and also builds your tannin profile. And this way I have absolute control." Haisma is working on this in Burgundy, with some interesting results, but he is unaware of anyone else doing it this way.

"Whole-bunch for me is about controlling the ferment, slowing it down, with a slow release of sugar," says Haisma. "It is a great way to build loads of complexity and savory characters, and still keep a lush creamy feel to the palate. I think of velvet. This is especially noticeable with my Cornas. As for burgundy, it's all about the complexity and finesse. In the big appellations I feel it adds a structure to the fruit, without adding coarseness or bitterness, characters I hate in Pinot Noir."

The other variable here is the length of time the stems stay in the ferment. Is a cold soak employed, or a post-ferment maceration? This could increase the extraction of flavor compounds from the stems.

Dirk Niepoort from Portugal's Douro uses stems to make his Charme wine, which is known for its finesse and elegance. For him, the length of time the stems are macerating for in the *lagar* is critical. In one vintage he says that he misjudged a *lagar* by five hours, and that was enough for the wine to be excluded from the final blend of Charme.

Many winemakers, however, are advocates of entirely whole-bunch or destemmed fermentations, rather than combining both in the same ferment. "I used to do either a hundred percent or nothing," says Anthony Yount of Kinero Cellars in Paso Robles, California. "Then, for a period of time we did some partial whole-cluster. This came out of working through a drought where the tannins increased [smaller berry size] and the alcohols crept up and we were extracting more tannin, so we were just trying to get a little more juice and less tannin extraction. Where I am today is a hundred percent either whole-cluster or destemmed, but what I do differently is to combine the ferments at some point. I start them as two ferments, then I take the hundred-percent whole-cluster halfway through fermentation and press that off and add the juice to the destemmed fermentation. I'm looking to get the aromatic freshness and vibrancy from the whole-cluster fermentation, a little bit of the pH change, and I'm getting it off the stems before the alcohol levels creep up." He then has 50 percent more juice in the destemmed ferment, which means a lighter extraction from those skins.

NEGATIVE EFFECTS OF WHOLE-BUNCH?

We have discussed the positive aspects of whole-bunch fermentation. What about the negatives? Blair Walter says that he used to do one fermenter with just whole-bunches each year but has now given up. "For us it is too much," he says. "It is interesting, but the wine becomes too herbal—it is like a hessian [burlap] sack character." But he still uses smaller proportions of stems in many of his fermentations. "With stems, people expect the wines to become angular. I find the opposite. Destemmed wines taste more angular. A lot of people don't have the courage [to use stems]; they aren't willing to tolerate earthiness and herbal characters in the wine." Tom Carson finds that using too many stems gives his wines a mulchy, herbal character.

"Whole-bunch is a very important part of my ferments," says Mark Haisma. "But the stems need to be clean. Any mold and it really shows in your wine, worse than moldy fruit."

Greenness is the problem most often associated with stems. While there has been increased interest in the use of stems in red wines worldwide, one region stands out as an exception: Bordeaux. This is likely because the main Bordeaux varieties of Cabernet Sauvignon, Cabernet Franc, and Merlot all share a degree of greenness in their varietal flavor signature, something that most winemakers will seek to minimize, and won't want to risk exaggerating by including stems. However, the late Paul Pontallier at Château Margaux looked at the impact of stems in the course of the extensive in-house research program. This stem trial was with 2009 Cabernet Sauvignon from a plot that, in good years, makes it into the first wine. "We wanted to see how important it is to destem," recalled Pontallier in a presentation on this work shortly before he died. "Our tradition has been to almost totally destem. From the early 20th century at Margaux destemming was a standard procedure." He pointed out that some were suggesting that using some stems could

be a good thing. And on the other side, that some estates had become more fastidious about removing even the tiniest traces of stem. The destemming regime in practice at Margaux leaves some tiny pieces of stems in the ferment, such that 0.03–0.05 percent of the ferment is stems. In this trial, the standard Margaux destemming was compared with one-percent stem additions, and one-percent stem additions, but with the stems cut into tiny pieces. To Pontallier, the results from this trial were obvious. His view was that the destemmed approach produced the best wines, and the one-percent stems in pieces the worst. But he was still cautious about generalizing the result. "We shouldn't draw too general conclusions. For this wine I think destemming works, but for other plots, such as a rich wine with soft tannins, it might be different."

In California, Paul Draper at Ridge also avoids using stems with Cabernet Sauvignon. "We have never used stems with the Bordeaux varietals as we have more than enough tannin in any year," says Draper. "In addition, in our cool climate—which is as cool as Bordeaux during the growing season but with cooler nights and warmer days—we are sensitive to any green character which, of course, is a risk with stems." Draper also chooses not to use stems for his Zinfandel: "Though this is not as tannic as the Bordeaux varieties, it is well balanced without any additional structure." However, he does include the stems when he ferments the few tons of Petite Sirah that Ridge have at Lytton Springs. Draper thinks there is good reason that stems are most widely used for Pinot Noir. "Given that Pinot Noir has less tannin and fewer different kinds of tannin than virtually any other well-known variety, the use of stems when needed makes more sense."

But another perspective is that the green character in whole-bunches doesn't come from having green stems, but rather from working whole-bunch ferments too hard. Waiting for stems to lignify —turn brown—before using them in fermentation, is one of the common myths of the whole-bunch narrative. In most regions, this won't happen until the fruit is overripe. According to some winemakers I have spoken with, green stems, if they are handled gently, can be used without anxiety about green flavors in the wine.

Another issue with whole-bunch ferments can be volatile acidity. Lukas von Loggerenberg, a rising star of the South African wine scene, uses whole-bunch, but he doesn't want too much of the carbonic-maceration character from whole berries persisting in the ferment, and he is worried about any air pockets present acting as sites where volatile acidity can form. So, he stomps the fermenting bins by feet to get all the oxygen out of the bunches and break many of the berries, so once it starts fermenting there are no air bubbles in the cap. "This is what normally gives a lot of volatile acidity," he says. "I know a lot of people talk about whole-bunch taking away the acidity, but for me it gives perceived freshness on wines. The tannin ages beautifully as well."

CONCLUSIONS

There are many different ways of doing whole-bunch fermentations. Combine these different techniques with the variability in the state of ripeness of the stems, and it creates a complex matrix of factors liable to result in different flavors in the final wine. So, it is with some trepidation that I'm going to attempt to sum up the way that whole-cluster ferments affect the flavor of red wines.

The state of ripeness of the stems seems to be very important, and this is likely to be determined primarily by the vineyard site, with vintage variation playing a role. In some warmer regions with a shorter ripening time, the stems may still be very green at harvest and thus unsuitable for inclusion at all.

An element of carbonic maceration is an important part of whole-bunch ferments. The intracellular fermentation that occurs in any

intact berries will produce interesting aromatic elements, and the slow, gradual release of sugar into the ferment will change its dynamics. Together with this, the reduced temperature of whole-bunch ferments is likely to have some effect on the resulting wine, usually in a positive direction. There may also be some direct flavor input from the stem material to the wine, which can be both good and bad, depending on the state of the stems. And the slight rise in pH that occurs with whole-bunch may increase the susceptibility of the wine to *Brettanomyces*, but at the same time may improve the mouthfeel.

What are the benefits of whole-bunch? One is textural—it seems to deliver a textural smoothness or silkiness that is really attractive, especially in Pinot Noir. Along with this, the tannic structure may be increased. I find that young whole-bunch reds often have a grippy, spicy tannic edge that can sometimes be confused with the structural presence of new oak. Frequently cited as a benefit of whole-bunch is the enhanced aromatic expression of the wine, and it's common to find an elevated, sappy green, floral edge to the pronounced fruity aromas, which is really attractive. Freshness is another positive attribute associated with whole-bunch.

Done well, whole-cluster can help make wines that are more elegant than their totally destemmed counterparts. I would add that whole-bunch wines sometimes start out with distinctive flavors and aromas that can be a little surprising (tasting terms associated with whole-bunch include broccoli, soy sauce, compost, mulch, forest floor, herbal, green, black tea, cedar, menthol, cinnamon), but these often resolve nicely with time in bottle.

And even commentators, such as UK retailer and Burgundy expert Roy Richards, who used to be opposed to whole-bunch fermentation, are softening their attitudes. "I no longer have an ideological view on this question and understand that it is rather more complicated than I used to believe," says Richards. "As a disciple of the late

Henri Jayer, I followed his mantra that stalks led to green tannins and that new oak to creamy, soft ones. And it is true that in his time his wines stood out for their vibrancy and sensuality whereas those wines from more illustrious domaines seemed a little delicate and pasty alongside." Richards adds that, "he is doubtless turning in his grave, seeing his protégés, Jean-Nicolas Meo and Emmanuel Rouget experimenting with whole-bunch fermentation in his beloved Cros Parantoux." Richards thinks that this could in part be attributed to changing weather patterns. "Burgundy is no longer such a marginal climate," he says. "I can understand from the results I have seen that stalks lend finesse and some floral perfume to wines that might otherwise be a little butch, say Corton, Clos Vougeot, Pommard, and certain Morey Premier Crus."

It seems that the circle has turned. What was once regarded as an outmoded practice—including the stems in red-wine ferments—is now becoming a fashionable winemaking tool for those seeking elegance over power.

CARBONIC MACERATION

Normal alcoholic fermentation involves the conversion of sugars in grape must into alcohol and carbon dioxide via the activity of yeasts. Carbonic maceration is a technique that delays this process, and which allows a fermentation independent of yeasts to take place, involving glycolytic enzymes present in the grapes. Carbonic maceration was first studied in detail by Michael Flanzy in the 1930s. Flanzy was director of the INRA oenological station in Narbonne. The Rhône wines at the time had toughness to them, and he saw carbonic maceration as a potential solution, which he described in a paper in 1935 entitled *Nouvelle méthode de vinification*.

One region in particular has become synonymous with this technique: Beaujolais. These enzymatically catalyzed intracellular reactions, often described as "fermentations,"

begin when oxygen drops below 5 percent or carbon dioxide is above 50 percent. One of the main conversions that takes place is the degradation of malic acid to alcohol (malic acid forms pyruvate, which is decarboxylated to give acetaldehyde, which is then reduced to ethanol). The enzyme malic acid dehydrogenase produces ethanol, succinic acid, and aminobutyric acid. Another enzyme, grape alcohol dehydrogenase, converts sugar to ethanol and carbon dioxide. This process produces various compounds that are important for flavor (or which are flavor precursors). For example, extra amino acids are liberated from grape solids, which increase the nutrient status of the juice, and open up the potential for these amino acids to act as flavor precursors. The ethanol produced can esterify some grape components, and one ester produced this way, ethyl cinnamate, gives strawberry and raspberry aromas. Another compound that increases is benzaldehyde, which adds cherry/kirsch aromas. There is also the production of vinylbenzene, which smells of plastic and styrene.

In addition, during carbonic maceration, polyphenols from the skins diffuse into the pulp and juice, and one result is that color is extracted but with less tannin than conventional maceration. The length of maceration can vary, but as a guide, ten days of intracellular fermentation gives 0.5–2% alcohol, with 15–75 percent of the malic acid metabolized. There is a fall in acidity levels that can be quite significant, with titratable acidity (TA) declining by as much as 3.5g/l and pH increasing by up to 0.6 units. However, bear in mind that there would be some loss of acidity during the malolactic fermentation that almost always occurs after alcoholic fermentation in red wines.

CARBONIC MACERATION VARIATIONS

There are variations on the theme of carbonic maceration. Strict carbonic maceration would be when grapes are placed into a tank in intact whole-bunches, with some care so that no juice

Above Concrete fermentation tanks used for carbonic maceration at Marc Delienne's winery in Fleurie, Beaujolais. Concrete is becoming fashionable again, in large part to its thermal inertia, which keeps fermentation ticking over nicely without temperature peaks and troughs.

Above Matthieu Lapierre in his cellar. Matthieu's father, Marcel, was a pioneer of natural winemaking in the Beaujolais region back in the 1980s. Matthieu and his sister Camille are carrying on their father's work.

is released. Then the tank is flushed with carbon dioxide and sealed. The carbon dioxide addition needs repeating, because it dissolves in the juice in the berries (the big danger here is that the volatile acidity can get out of hand if oxygen is present). Then, maybe a week to ten days later, the tank is opened. The berries will still be intact and will have a slight prickle to them when tasted. The flesh will have absorbed color from the skins. The whole-bunches are then crushed, and a normal alcoholic fermentation finishes off what the carbonic process has started. But in most cases, carbonic maceration is not as strictly carbonic as this. There are many variations on the theme, but in Beaujolais the common method is to fill the tank or vat with whole-bunches, and then for some at the bottom to release juice under pressure from the grapes above them, and a normal alcoholic fermentation starts at the bottom, releasing carbon dioxide that then protects the bunches above that are undergoing carbonic maceration.

Marc Delienne is a natural winegrower in Fleurie, Beaujolais. He explains how he does two variations on carbonic maceration. The grapes are brought and whole-bunches are put into the tanks. "We do nothing chemical," he says, with

his only winemaking addition being some sulfites before bottling. He has two different processes with his Gamay grapes: pure carbonic maceration and semi-carbonic maceration. "For me, semi-carbonic means I put grapes in and most days make a short pump over." There is no punching down or foot treading. He adds that for semi-carbonic he adds perhaps 500kg (about half a ton) of destemmed grapes at the bottom of the vat to get an alcoholic fermentation going. The duration of fermentation in the tanks is between 6 and 21 days. It depends on the vintage and a lot of other things, but essentially his taste. He makes five different Fleuries: three in semi-carbonic, and two in full carbonic maceration. "My carbonic maceration is an extreme carbonic maceration," says Delienne. This is because all the time he leaves the valve at the bottom of the tank open. Any juice produced by pressure on bunches at the bottom of the tank is removed, and then may be used later on in this or another cuvée. "The fermentation is totally enzymatic in the berry," he says. "Every day I take grapes and I taste, I see the color, I taste the tannins, and when it is okay for me, I press, and the yeast can work." The maceration temperature is kept between 20 and 25°C (68–77°F).

On a visit to Lapierre, a well-known natural winery in Morgon, Beaujolais, I asked Matthieu Lapierre to explain his winemaking approach. "We need the grapes to be as perfect as possible to go into the vats," says Lapierre, "because we do carbonic vinification." We go into the winery and see some large wooden vats. I ask whether this is where he does fermentation. "The maceration is not really a fermentation," says Lapierre. "These vats are only used during vinification time to do the carbonic. We fill them with good grapes, but we want the grapes to be fresh. If the weather is too hot, we rent some cooling trucks, and it allows us to chill the grapes overnight." He puts three to four tons of grapes in each vat, and with the weight some grapes are crushed at the bottom. "I call it the cooking juice," he says, "because that is what it is. The juice will provide the carbon dioxide by fermentation with indigenous yeasts, and this is necessary to store the grapes in an air-free atmosphere. We have more or less juice depending on the vintage. If we have too much juice, we can take it out, to get the right level. But the best way to control this is the way you put the grapes in." If the grapes are more delicate, they are put in slowly, so as not to break them. "If the vintage gives grapes that are harder, it is better to use a conveyer, so you have enough juice." He thinks they need five–ten percent juice for a good maceration. "The grapes inside the vat don't ferment at all. What happens is that enzymes inside the fruit start to attack the cells, making the color move from the skin to the inside of the berry. At this time a lot of acids are being decomposed by the enzymes, and some basal flavor is being created. So, without fermentation you will have some aromas and color extraction, and also some biological transformation of some components into others. This creates some of the base of flavor of the vintage."

"We press the grapes after two, three, or four weeks," he says. "In 2019 the average was 19 days; not that long." By the time the grapes are pressed they are still intact, but are darker and a little bit more red. They are whole-cluster still, and so it is quite a challenge to get them into the press, which resembles a Champagne press because of its shallow cage and wide diameter. This leaves behind the skins, stems, and seeds. "At the end everything is just flat," says Matthieu. "The berries are still connected to the stems. We have just extracted the juice, which is red but full of sugar." The fermentation starts when they add the free-run juice, which already has lots of yeast in it. "Then, as quickly as possible, we barrel the wine." The juice has some properties of wine from the carbonic, but most of the alcoholic fermentation is therefore off the skins. This is one of the key aspects of carbonic maceration: because there is no maceration of the skins in the presence of alcohol, the extraction is light, and the wines resulting often have quite low tannins and color, and a distinctive fruity aroma.

Once fermentation has finished they look at the wine through a microscope to see what is there. "It is always the same question," says Lapierre. "Do I need to sulfite at the end of fermentation or not? This is the most critical point. The degrees [of alcohol] get higher so the natural yeast has less activity, so they are giving more space to the bacteria. Now we have 2 or 3g (0.07–0.1oz) of sugar left. If there are not a lot of bacteria we can say we are stable enough." If there is too much sugar and they spot the wrong sort of bacteria, "sometimes it's better to do something."

His father, Marcel, began looking at the fermentations under the microscope, after learning what to look for with the help of a microbiologist. The other natural-leaning Beaujolais producers also tend to do this, and Lapierre shares his microscope and know-how (they have a microbiologist who assists to this day) with friends.

Lapierre showed me the microscope room. In one wine sample, he points out that there is quite a bit of *Saccharomyces cerevisiae* present, which is good. They are budding, so they are alive and healthy. But there are also quite a lot of bacteria, as single cells, as pairs, and as chains. If these eat up the sugar, then the volatile acidity will rise. And if there's any *Brettanomyces* there, they can have a feed too, with bad results.

So, what would he do with a wine like this? First of all, he'd try chilling it down to about 14°C (57°F), which will allow the yeast to carry on, but which is a little too cool for the bacteria to cause much trouble. And he would do some *battonage*, stirring up all the yeasts, alive and dead. If the sugar doesn't start dropping, then he would consider adding some sulfur dioxide: enough to knock out the bacteria, but not enough to take out the yeasts. "It's a race," he says, "and the horse that we want to see win is this one," pointing at *S. cerevisiae* on the screen. Lapierre likes to emphasize that how they work is not just about not using sulfites. It is about doing a natural ferment. He will use sulfites if he has to and nothing else works; 2015 was a tricky vintage because of the ripeness levels, and there were lots of problems completing the ferments. "We didn't let the brett ferment the wine in 2015," he says. "We did sterile filtration where there was brett and reintegrated them with barrels that were fermenting with interesting yeast."

13 Sulfur dioxide

Sulfur dioxide (SO_2) in wine is one of the most frequently discussed and yet simultaneously one of the most frequently misunderstood issues in winemaking. It is an important subject, so it is a good idea to have a decent grasp of the issues relating to its use. First, I will look at why SO_2 is such an important component of winemaking, and how it acts as a "chemical custodian" of wine quality. Then I will turn to the important subject of how SO_2 can best be used, and why it is generally not a good idea to use too much or too little of it. Finally, and perhaps most interestingly, I will report on attempts to make sound wines with no added sulfites.

WHY SULFUR DIOXIDE IS JUST ABOUT INDISPENSABLE FOR WINEMAKING

Sulfur dioxide acts as a guardian of wine quality in two ways. First, and most importantly, it protects the wine from the ill effects of oxidation. Secondly, it acts as an antimicrobial agent, preventing the growth of unwanted spoilage bugs in the wine.

Peter Godden of the AWRI (Australian Wine Research Institute) describes SO_2 as a "magical substance" because it has these effects at very low concentrations. "We're mostly talking about a maximum of one hundred and fifty parts per million (150mg/l) and much less in most Australian wines," he explains (Godden showed data indicating that of a representative sample of Australian wines, at least 70 percent have levels below 100mg/l, and close to 60 percent have 80mg/l or less). Correct use of SO_2 is a subject that he and his colleagues of the AWRI Industry Services team have been spending a lot of time advising Australian winemakers on in recent years. "A wide range of the problems that we see have relatively few root causes, and SO_2 use, or misuse, has been the most common."

FREE AND BOUND

Key to understanding the effects of SO_2 is the ratio between the free and bound forms. When SO_2 is added to a wine, it dissolves and some reacts with other chemical components in the wine to become "bound." This bound fraction is effectively lost to the winemaker, at least temporarily, because it has insignificant antioxidant and antimicrobial properties. Various compounds present in the wine, such as ethanal, ketonic acids, sugars, and dicarbonyl group molecules, are responsible for this. Winemakers routinely measure total SO_2 and free SO_2, with the difference between the two being the amount in the bound form. Importantly, equilibrium exists between the free and bound forms such that as free SO_2 is used up, some more may be released from the bound fraction.

It is actually slightly more complicated than this explanation, though, because some of the bound SO_2 is locked in irreversibly and the remainder is releasable, and of the free portion most exists as the relatively inactive bisulfite anion (HSO_3^-) with just a small amount left as active molecular SO_2.

From the winemaker's perspective, molecular SO_2 is the interesting part. Typically, levels of 0.8mg/l molecular SO_2 will be aimed for in white wines, and this will probably require there to be 15–40mg/l of free SO_2 present. Reds can get by with a little less.

THE IMPORTANCE OF PH

One of the key factors affecting the function of SO_2 is pH. This is a measure of how acidic or alkaline a solution is (technically it relates to the concentration of hydrogen ions in solution), where a pH of 7 is neutral and below and above this the solution is progressively more acidic or alkaline, respectively. Thus, a wine with a lower pH is more acidic. All wines are acidic (with a pH of less than 7), but some are more acidic than others. In two respects, pH is important here. First, at higher pH levels more total SO_2 is needed to get the same level of free SO_2. Second, SO_2 is more effective, that is, actually works better, at lower pH. So, as well as having more of the useful free form for the same addition, what you have works better as well. It is a double benefit. This is shown in the table below.

Percentage of free sulfur dioxide in the molecular form at different pH levels

pH	percentage molecular SO_2
2.9	7.5
3.0	6.1
3.1	4.9
3.2	3.9
3.3	3.1
3.4	2.5
3.5	2.0
3.6	1.6
3.7	1.3
3.8	1.0
3.9	0.8

The most useful attribute of SO_2 is that it protects wine against the effects of oxidation. There are two sorts of oxidation processes in wine. The first, which only really happens in grape must, is enzymatic, and is caused by oxidases. The levels of these enzymes are much increased in damaged or rotten grapes, so where these are likely to be present it is especially important to use sufficient SO_2. The message here for winemakers is that it is important to use grapes that are as clean as possible, with the absolute minimum of fungal damage.

The other types of oxidation reactions that occur in wine are chemical. Oxygen itself isn't terribly reactive, but it is made reactive by the presence of reduced transition metal ions, principally iron and copper. The next stage is the production of quinones from phenolic groups, and also the production of hydrogen peroxide, a powerful oxidizing agent. While SO_2 does not react with oxygen itself, if it is present in the free form it can bind with the quinones and peroxide and take them out of the equation. Otherwise they would go on to react with other wine components. Sulfur dioxide can also bind with the products of oxidation such as ethanal (also known as acetaldehyde), which otherwise would make the wine taste and smell oxidized. So, it isn't preventing oxygen from having an affect on the wine, but it is limiting the damage and clearing up some of the mess.

White wines generally need higher levels of SO_2 than reds to protect them. This is because red wines are richer in polyphenolic compounds, which give the wine a natural level of defense against oxidation. White wines that have been handled reductively (that is, protected against oxygen exposure through the use of stainless steel and inert gases in the winemaking process) are especially vulnerable to oxidation and need careful protecting.

Sulfur dioxide is also microbicidal. It prevents the growth—and at high enough concentrations kills—fungi (yeasts) and bacteria. Usefully, SO_2 is more active against bacteria than yeasts, and so by getting the concentration right winemakers can inhibit growth of bad bugs while allowing good yeasts to do their work. It is quite common for winemakers to add 50mg/l (the same as ppm) SO_2 to the grapes on reception, to knock back wild yeasts and tip things in favor of *Saccharomyces cerevisiae*. It is usually still added to the crushed grapes in wild-yeast fermentations. While it kills some of the natural yeasts present on grape skins, the stronger strains survive and thus are selected for preferentially.

BEST PRACTICE IN SULFUR DIOXIDE USAGE: GETTING THE RATIO RIGHT

But while the microbicidal action of SO_2 might encourage some winemakers to add in more just to be on the safe side, Godden suggests that the best way to ensure wine quality is not by using more SO_2, but by using it smarter. His idea is that the key measurement for winemakers is not the free SO_2 level, but the ratio of free to bound SO_2. That is, the key to effective SO_2 usage is getting the ratio of free to bound SO_2 as high as possible, to maximize the benefits of the amount added. He sent me data gathered by the AWRI Analytical Service on a typical cross-section of Australian wines that show that in a range of reds, free SO_2 has been steadily increasing in recent years while the total SO_2 has actually been decreasing. Thus the ratio of free to total SO_2 has improved. "I consider the use of the ratio of free to total SO_2 as one of the most useful quality-control measures during winemaking," says Godden.

Above Adding SO_2 (as potassium metabisulfite solution) to a fermenter: SO_2 is almost universally added to wine as a microbicide and an antioxidant (though it is not strictly an antioxidant), but its levels are tightly regulated.

How is a good ratio achieved? Starting with healthy grapes is important. Grapes suffering from rot have significantly higher levels of compounds that will bind SO_2 and also enzymes that encourage oxidation. Judicious filtration, where necessary, will also help make SO_2 additions more effective, by reducing microbial populations to a level where the SO_2 is more effective against them. General cleanliness in the winery is also helpful.

Perhaps most important, though, are two critical winemaking interventions: first, controlling turbidity by careful racking, fining, and filtration (if necessary); and second, the timing and size of additions. There are three points where wine is likely to be subject to considerable oxygen stress or risk of bug growth: at crushing; at the end of malolactic fermentation [MLF] (or alcoholic fermentation where malolactic is discouraged); and at bottling. At each of these points a healthy dose of SO_2 is highly recommended. Crucially, for the same total addition, it is much more effective to add your SO_2 in fewer relatively large amounts rather than many small additions. In the latter case you run the risk of never getting your free SO_2 levels high enough for it to do its job properly.

HEALTH EFFECTS

Now we turn to the consumer's perspective. Is SO_2 healthy? Not completely, is the simple answer. Sulfur dioxide can cause adverse reactions in some asthmatics, which can be quite dangerous at ingestion levels as low as 1mg. For this reason some doctors have even gone as far as suggesting that asthmatics avoid wine altogether. For most people, it is probably fairly harmless at the levels used in winemaking, but anyone drinking wine on a regular basis is probably taking in more than medical experts recommend (although it could be debated that these levels are set a little low, in the name of caution). Sulfur dioxide levels in wine are subject to regulation by various authorities. The EU has set a maximum permitted level that varies with wine type from 160mg/l (or parts per million) for dry red wines to 300mg/l for sweet whites and 400mg/l for botrytized wines. In Australia, the regulations permit 250mg/l for dry wines and 350mg/l for those with more than 35g/l of residual sugar. In the USA, the maximum level allowed is similar. Any wine with more than 10mg/l SO_2 (this level can be reached naturally even if no SO_2 is added) has to be labeled "contains sulfites."

On the basis of animal experiments, the WHO (World Health Organization) has set the RDA (recommended daily allowance) of SO_2 at 0.7mg/kg bodyweight. Doing some simple sums, this would permit a 70-kg (154-lb) human to take in 49mg SO_2/day. Half a bottle of wine with an SO_2 level of 150mg/l would provide 56mg of SO_2, thus exceeding the recommended daily allowance.

Many people who suffer adverse reactions to wine, such as headaches and flushing, blame SO_2, partly, one suspects, because it seems an obvious candidate as an added chemical substance. After all, the bottle will have "Contains Sulfites" written on it, usually with no other ingredient labeling. The issue of adverse wine reactions is a complex one, and the scientific literature offers few clear indications of the culprit compounds. However, many foodstuffs contain higher levels of sulfites than wine, with the worst offenders being dried fruits, which typically contain about ten times the level in wine.

"VINS SANS SOUFRE": THE QUEST FOR NATURAL WINE

So, if the presence of SO_2 is just about essential for winemaking, why would anyone want to do without it? There are two reasons. First, people have increasing concerns about what they put into their bodies, and so are anxious not to consume anything that has been chemically manipulated. To many people unaware of the issues, SO_2 use sounds like a gratuitous addition of unnecessary chemicals. This creates a potential market for "additive-free" wines. Second, there exists a band of passionate winemakers who see wine as a natural product. Eliminating SO_2 usage is seen as the final hurdle in the quest for fully "natural" wines. The desire for naturalness runs strong, and it's not just fringe winemakers who pursue this goal.

When I wrote the first edition of this book in 2004, the natural-wine movement was a tiny group, with just a few way-out winegrowers working without added SO_2. Now, 15 years later, it has grown into a sizeable, lively coalition of hundreds of winegrowers. And while many do add a little SO_2 at bottling, there are many who avoid this additive altogether.

This market really began in the 1980s with a circuit of wine bars in Paris, all of which wanted to serve fresh wines with a purity of fruit to them. An instrumental figure in this trend has been Jacques Néauport, whose inspiration was the late Jules Chauvet. Chauvet was a small négociant in the Beaujolais region with an enquiring mind who tried out a number of novel ideas, one of which was making wine without added sulfites. Since Chauvet's death, Néauport has consulted for a number of growers, the first of which was Overnoy in the Jura, and which included the late Marcel Lapierre in Morgon. Néauport developed a vinification method specially adapted for working without SO_2, involving carbonic maceration under very cold conditions. Catherine and Pierre Breton, Thierry Allemand, Jean Foillard, and Pierre Frick are other well-known pioneers of working without SO_2 for at least some of their wines. A common claim is that wines made in this way have a greater purity of fruit and are aromatically more interesting.

It must be pointed out, though, that even if no SO_2 is added during the winemaking process, there will still be some present in the wine because it's a byproduct of fermentation. Yeasts produce small quantities or around 5–15 mg/l quite naturally, so the notion of a totally sulfite-free wine is illusory. Dominique Deltiel, a consultant winemaker based in France, did a survey of some yeast strains and found quite a few that produced larger amounts of SO$_2$ during fermentation, including some making as much as 100mg/l.

But it is not just the natural wine community who are advocating the use of fewer sulfites in wine. Arthur O'Connor, of Disruption Wine in California, first started making wine in California in the late 1990s with Bonny Doon. It was a time of change in the vineyards and the wineries. Everyone was worried about *Brettanomyces* and started using higher levels of SO$_2$. One winery he worked with even acidified using tartaric acid to lower the pH in order to have more molecular SO$_2$, and then prior to bottling they deacidified. Red winemaking was moving from oxidative to anaerobic, including the use of a cold soak in the presence of SO$_2$. The effect was that the SO$_2$ was locking in all the polyphenolics very early on. "The result was some very green wines," he recalls.

Recently, some larger companies have begun making larger quantities of wines without added sulfites. Most notable among these efforts is the Naturae range from Languedoc producer Gérard Bertrand. They make three million bottles of these affordable wines without adding sulfites. I recently (July 2020) tasted through the range, right back to the first wines made in 2011, and they are all showing really well, which might be a surprise considering they have no protection from sulfites. I quizzed winemaker Stéphane Queralt, who is the Gérard Bertrand winemaker for partnerships, about his approach. "To produce Naturae we have a close partnership with the winegrowers and we do a selection of the parcels. We select three times: the first is at the end of spring, then at the beginning of summer, and if the parcel has good quality and the grapes are top quality for making Naturae, we decide to harvest them when they have good acidity and sugar. We do the selection of the parcel, control the harvest, and do the vinification."

He has a team of 12 winemakers who are working in the partnership cellars. One of their partner wineries is Caveau d'Héraclès, which is the largest organic wine cooperative in France. Located in Vergèze in the Gard, it makes seven million liters (about 1.85 million gallons) of organic wine a year. For the whites they do hyperoxidation of the must, which he says is very important, getting rid of the phenolics. The phenolics oxidize, the must turns brown, but then after fermentation the wines are much more stable. They cold settle but not too much: they want to take a bit of sediment because they are not allowed to use nutrients. Fermentation is not too cold, because this produces sulfides, and so they aim at 18–22°C (64–72°F).

One of the techniques they practice is known as bioprotection, something consultant winemaker Stéphane Yerle helped them with. The idea is to use biology to create a competitive environment, which means that the bad microbes have no space to thrive. Yerle calls his approach "active bioprotection." The main principle is competitive exclusion, by using a range of microbial inoculants that populate the must and then the developing wine, making fermentation more secure. But there are also other benefits. Removing sulfites helps avoid producing aggressive red wines with excessive tannin extraction. By replacing the SO$_2$ with microorganisms that occupy the space in the wine and produce interesting metabolites, it diminishes the excessive solvent nature of the juice and results in a gentler extraction. In addition, *Lachancea thermotolerans* is used, a non-*Saccharomyces* yeast that produces some succinic acid and lactic acid, adding a sense of freshness to the wine and increasing the acidity.

Teresa Garde Ceirán described some experiments on bioprotection that her group at the ICVV experimental winery in Logroño (Rioja, Spain) have carried out. They looked at Tempranillo: 400kg (0.4 tons) of grapes were divided into four batches, with duplicate 100-liter (26-gallon) fermentations. The control treatment was inoculated with a selected strain of *Saccharomyces cerevisiae*, and the bioprotection treatment was initially inoculated with a malolactic bacteria *Lactobacillus plantarum* (Viniflora NoVA, CHR Hansen), then 48 hours later inoculated with a non-*Saccharomyces* yeast *L. thermotolerans* (Vinflora Concerto, CHR Hansen) at a controlled temperature of 18°C (64°F). This will do some of the fermentation and then *S. cerevisiae* will take over. After alcoholic fermentation finished, the wines were pressed and the control was inoculated with malolactic bacteria *Oenococcus oeni* CH16. The bioprotection wines were inoculated with the same bacteria because there was some malic acid left. They analyzed the wines and did sensory work. Fermentation kinetics were faster in the control wines. *L. plantarum* consumed malic acid for the first three days, and then it stayed at a constant level for eight days, and then the rest of the malic acid was consumed by the inoculated bacteria. There were no effects on the color of the red wines. The bioprotected wines showed more aroma intensity, and analysis showed they had more aromatic compounds. They got good scores by the tasters; there were no significant differences in the flavor, but they were rated more fruity and fresher.

Queralt says that when fermentation is done, they do an inoculated MLF, so that the wine is then stable. It is then bottled, using nitrogen to protect from oxygen pick-up. Pick-up is 0.2mg/l, which is very low. This helps protect the wine.

For the red, where fermentation is finished they do some fining after MLF, at cold temperature. This is to help take out any *Brettanomyces* cells that might be present. Then they do an analysis of the bacteria or any other microbes present. "This is very effective against *Brettanomyces*," says Queralt.

The key is to do the MLF straight after alcoholic fermentation, in a short time. He has tried co-inoculation, but they have run into problems with mousiness. They always find less than 10ppm in their wines, so they are clearly using a strain of yeast that is a low sulfite producer. They sterile filter the wines using a plate filter.

"What is interesting is that the old vintages of sulfite-free wines are fresher than some of the wines with sulfites," says Queralt. Are sulfites somehow encouraging oxidation? "Yes, if you don't bottle carefully and put oxygen in your bottle, the sulfites combine with the oxygen. After that you have ethanal, and you still have the ethanal in the wine. The ethanal taste makes the wines tired, because there are no sulfites left to protect."

Daniel Dycus of Laffort USA (a yeast and microbe company) describes some experiments that they have been involved with at the UC Davis. In one, they looked at the bioprotective effect of non-*Saccharomyces* yeasts on compromised Cabernet Sauvignon fruit (it had sour rot and had been kept three days in a cold room). In the control, 60ppm SO_2 was added and fermentation was allowed to take place. Initially, there was a big population of wild yeast *Hansenosporia*, which is sensitive to alcohol but not sulfites. This produced a lot of volatile acidity. But if the must is inoculated with a mix of *Metschniakowia* and *Torulospora*, they outcompete the *Hansenosporia*, and this is more effective than SO_2 in protecting the wine. Dycus also describes an innovation called "the living harvester." Machine harvesters are covered in sugar so make an ideal breeding ground for bacteria and yeasts. Laffort devised a yeast gun to spray a mix of strains all over harvester surfaces in a five-minute protocol to outcompete any bad microbes.

Finally, a wide-spectrum antimicrobial called chitosan is increasingly being used to deal with *Brettanomyces* problems. This comes from the deacetylation of chitin sourced from the fungus *Aspergillus niger*. At wine pH, this is positively charged. Any live cells in the wine have a negative charge and these adhere to the chitosan. The complexes that form precipitate and float to the bottom of the tank. The chitosan interrupts the metabolism of *Brettanomyces* and the cells lyse and die. If a wine has a brett contamination, you can use higher doses of chitosan and then rack it off its lees. There is no single SO_2 alternative that does everything that SO_2 does, but these tools used together make working without sulfites a lot safer.

14 Wine faults: where are we, and when is a fault a fault?

In 2018, I published *Flawless* (University of California Press, Berkeley), a book on wine faults. In this chapter I give a quick summary of some of the ideas from this book, outline some of the critical concepts behind the concept of wine faults, and also take a quick drive-by look at some of them individually. It is a really interesting topic, even though the notion of faults does seem a little negative. Here, I am using the term "fault" to refer to both faults and taints. A taint is something from outside that then affects the wine negatively, whereas a fault is something that develops during winemaking. Some people consider them separately.

Let's begin by exploring some background concepts. First, beauty is not the absence of flaws. We have this notion that if we simply eradicate all flaws, then the result will be something beautiful. It is not as simple as this. Flaws can get in the way of beauty, but they can also be part of the beauty. The Japanese concept of *wabi sabi* is important here. It is hard to explain briefly, but the essence of this notion is that small flaws can bring out or enhance beauty. Think about a small drop of water falling into a pool, a flaw that brings out or emphasizes the quality of silence. In the context of wine, small traces of what might otherwise be faults at higher levels can bring out the beauty of the wine.

Some wine faults are always faults. Think of cork taint. If you can detect it, it is a problem. Others are context dependent. If you are a winemaker, then part of your training is to learn to spot wine faults very early on in the winemaking process. Baby wines with potential problems need to be identified early, so that something

can be done to remedy the problem, or the wine can be removed from the cellar. This is probably why winemakers often become fault policemen. They are only doing their job: they are like studio engineers who are recording music. Their interest is not so much in the song that's being played, but in the technique of recording all the parts as well as possible, and then putting them together in a mix. It is easy to be distracted from the main task—the actual output in terms of the finished wine—and get a little obsessed by the components.

I admire the skill of winemakers, but sometimes it can be a bit frustrating working with winemaker-led judging panels, and the results can often disappoint. Many of the most interesting wines get pushed back because of the detection of some miniscule component identified as a fault. And once someone on a judging panel has spotted a problem, it is very hard to disagree, in case it looks like you have missed something that should be obvious, thereby marking you as a poor judge.

My point: the presence of a fault compound does not make a wine faulty. Just because someone has recognized reduction or *Brettanomyces* (brett) in a wine (we will come to these specific faults shortly), it doesn't mean that we have to reject the wine as flawed. It is all about seeing the wine as a whole. To use the studio engineer analogy, it is about being able to listen to the song, not zeroing in on each track or assessing the production values.

For the last few years I have been working as one of the six co-chairs for the International Wine Challenge. This is one of the largest, most rigorous international wine competitions, and each year some 15,000 wines from around the globe are entered. Judging is by panels of five, usually with a range of skill sets being represented on each panel. Judging is a two-stage process, with everything being tasted in the first round, and then a proportion of those wines going through to round two to potentially be awarded medals. This means a lot of bottles are opened.

Around a decade ago, Sam Harrop MW, a consultant winemaker from New Zealand, initiated a faults clinic as part of his role as one of the co-chairs. All the wines from the panels that were considered to be faulty ended up on a separate table, and Harrop tasted them, and decided whether or not they really were faulty and, if so, which was the primary fault. When Harrop left, I took over this role, and it has been an interesting experience.

We now have ten years of data on faults in the real world, and a few years ago the Australian Wine Research Institute (AWRI) became involved, doing some chemical analysis of specific faults and interpreting the data that had been collected to date. At some stage the results will be published, so I can't talk about them in any detail. However, a preliminary report has already been published taking a broader look at some of the trends. From my perspective, I can report that faults are not

Above Sniffing through a series of doctored wines that have had different fault compounds added to them at the AWRI in Adelaide, Australia. Although it's very hard to replicate real-world faults in this way (they are often multifaceted), this is a useful excercise.

always clear cut. Sometimes, wines are simply not very good, rather than possessing an actual, specific fault. Other wines have more than one fault. And even professionals struggle with some faults, misclassifying them, or coming up with false positives. One of the problems is that it is very hard to train people on faults. You can add fault compounds to wines to create standards, and there is some merit to this, but faults express themselves differently in different wines. We just need to maintain a bit of humility here. Having said this, humans are quite good at smelling and tasting some things, even at very low levels. Let us take a look at some specific faults.

CORK TAINT

First, cork taint. It is still with us! Despite all the advances made by cork companies, and the new techniques for cleaning cork, this is an ongoing problem. As a natural product—tree bark—it harbors microbial communities. Pores called lenticels run through the bark, and these will always have microbes living in them. And with this microbial habitation comes the possibility of musty taint, chiefly caused by the presence of the haloanisole 2,4,6-trichloroanisole (TCA). When the bark is sorted, there are sometimes discolored areas that need to be rejected: these can also be a source of microbes that create taint compounds.

There is a lot of discussion about the rates of cork taint, and some good evidence that in the past it was alarmingly high, at over five percent, perhaps even ten. The problem is that people differ in their thresholds for TCA, and it also seems that low-level TCA has an effect on wine that strips it of character without imparting obvious mustiness. And there are also environmental sources of musty taint, so it would be wrong to assume that all wines with mustiness are "corked." Having said this, very few wines have musty defects that are sealed by screwcaps or other noncork closures, so most mustiness in wine does seem to be cork derived. In the 1990s, there was a massive increase in demand for cork, caused by the expansion of the global wine industry, and the lack of alternative closures. Some speculate that this large expansion led to a dip in quality and increased levels of cork taint. It was bad enough that the Australians and New Zealanders switched almost entirely from cork to screwcap in the early 2000s, which was in effect a brave experiment. But it worked quite spectacularly. Now, cork taint seems to be hovering between one–three percent of bottles sealed with corks. It is still too high, of course, but it is better than it used to be. There are alternatives to cork that have taken a large market share. Screwcaps are well established and come with a number of different liners offering different levels of oxygen transmission. DIAM is a technical cork made of small granules of cork combined with food-grade glue and plastic microparticles. The cleaning process, using supercritical carbon dioxide (a combination of pressure and temperature that gives it the cleaning power of a liquid but the penetration ability of a gas), results in zero taint, and the closure looks and feels pretty much like a cork. It comes in a range of composition densities giving a choice of levels of oxygen transmission. Other cork companies offer these technical corks— the term used to describe cork-based closures made of smaller fragments of cork glued together—which have been cleaned in a variety of ways so that cork taint risk is vastly reduced. And then there is Nomacorc, which is the dominant synthetic "cork." In the past, plastic corks used to be highly variable, but Nomacorc's latest products look and feel much nicer. Their SelectBio series has a significant amount of plant-derived plastic in it.

These alternatives have taken a good chunk of the market for cork, and the result is that far fewer closure-tainted bottles now reach consumers. An average person is quite good at detecting TCA, and we pick it up at such low levels that it is tough to analyze a wine for it. You need to use a technique called gas chromatography-mass spectrometry (GC-MS), and this is not a standard lab analysis.

BRETTANOMYCES

Brettanomyces is one of the most interesting and highly discussed wine faults. Brett, as it is known, is the presence of metabolites produced by a yeast called *Brettanomyces bruxellensis,* which give wine a distinctive smell and taste. This yeast is present in the wild, and tends to hang around until fermentation has finished, when it starts feeding on things *Saccharomyces cerevisiae* has left, or which *S. cerevisiae* cannot eat. As it feeds and grows it pumps out some flavor chemicals, chief among which are the volatile phenols 4-ethylphenol and 4-ethylguaiacol. It is mainly a red wine problem, too, as whites tend to have lower pH, which is protective, and not as many of the compounds that brett likes to feed on.

Even professionals can struggle to identify brett, because it varies in its expression. At the highest levels it tends to make wines smell quite stinky, like animal sheds and horse saddles. It gives a savory spiciness to wines, and sometimes they smell medicinal, like adhesive bandages made of fabric or antiseptic ointments containing phenol. Brett strips out the unfermentable residual sugars in red wines, so it can change their texture and make them seem drier. It also frequently gives a metallic edge to the wine.

Most consumers would not be able to pick out brett in a wine, but they might reject the wine without realizing why they are doing it. I certainly remember encountering this meaty, animal-like, medicinal character in some reds and deciding I didn't like them long before anyone told me what this was. Brett's varied expressions depend on the strain, the level at which it is present, and the wine characteristics. Sometimes it seems to work: there are a few wines that I can identify as being somewhat bretty, but which taste really nice. I would not want to have brett in my wine if I were a winemaker, but some people are lucky and seem to get away with it. This makes it a very interesting fault. Most winemakers I know would reject a wine immediately if they detected any brett.

OXIDATION

It is worth thinking of oxidation alongside another wine fault, volatile acidity (VA). They are close but different, and often occur together. Oxygen is necessary in winemaking, but as well as being a friend it can also be a foe. It helps to think of oxygen exposure in terms of macro, micro, and nano in the life of a wine. During fermentation oxygen is quite welcome as it helps it proceed smoothly. This is the macro. Then, depending on wine style, some oxygen—but much less—is helpful for the development of the wine (this is chiefly why oak barrels are used). Sometimes oxygen is deliberately introduced (this is microoxygenation). Then, later in the wine's life it should be protected more from oxygen, with perhaps just a very little transmission through the closure. This represents nano-oxygenation.

Oxygen management is a key issue for wine quality. And the appropriate exposure to oxygen differs with wine style. Reductive winemaking with an almost total absence of oxygen is right for some wines (bright aromatic whites, for example), but would be disastrous for others (big tannic reds). Managing oxygen pickup during wine movements and bottling is important for wine quality. And the appropriate use of sulfur dioxide can help remove the product of oxidation reactions and thus fight oxidation (see chapter 13, page 165).

Volatile acidity is usually a two-step problem. It is produced by acetic acid bacteria (and also some malolactic bacteria), but these need oxygen in order to grow. Red wines are particularly susceptible because they are fermented on their skins, and if the cap is not kept wet then this provides an ideal environment for these bacteria to do their work. The best way to counter volatility is having clean grapes coming into the winery (some musts will already have significant levels of acetic acid because of problems in the vineyard), and then controlling the oxygen exposure of the developing wine. Most wines have some VA, and this gradually creeps up during the winemaking process. The idea is to prevent this reaching problem levels. At low levels VA can add a pleasant lift to a wine's aroma; as these get higher, it becomes off-putting.

Some wines are oxidative in style, so it is not possible to say precisely when a wine is oxidized and when it is just a style choice. A lot depends on the intent of the winemaker, and the market for the wine. And older wines tend to be more oxidative than younger ones. When is a wine too old? People differ in their preferences. Of course, we also have some wine styles that are by all accounts fully oxidized: Oloroso Sherry, Madeira, and old Tawny Port.

REDUCTION

Reduction is one of the most interesting faults, but the name is a misnomer. In a redox reaction, oxidation and reduction are opposites. A chemical entity is reduced while another is oxidized, as they exchange electrons. But reduction in wine is not the opposite of oxidation, and it is perfectly possible to have a wine that is, at the same time, both oxidized and reduced. Maybe we should stop using the term "reduced" in the context of wine, because even some professionals fail to understand this distinction. Reduction is all about volatile sulfur compounds (VSC) produced by yeasts during fermentation, and then the way in which these compounds change form after fermentation has finished. The level of production of these compounds depends on the yeast strain, the nutrient status of the must, and also any stresses that occur, such as sudden temperature changes.

Bad reduction is horrible: hydrogen sulfide is the gas that smells of rotten eggs and drains, and during fermentation yeasts can often begin producing this eggy smell. Often, it will simply be a transient state and the wine will be fine. Sometimes a racking off the dirty lees can do the trick. But if it persists in the wine, it is a problem. More complex sulfides and related compounds such as mercaptans (thiols) can give a range of aromas, and also change the mouthfeel of the wine.

Sometimes, in the right context, VSC can be appealing. There is the famous example of matchstick/struck flint reduction, which many winemakers look for in white wines, especially Chardonnay. This is thought to be caused by a mercaptan called methional. Then there are some roast-coffee characters from barrel fermentation that can be attractive, and also the range of sulfur compounds that are important in the aromas of traditional-method sparkling wines.

A great example of positive VSC is the presence of polyfunctional thiols in Sauvignon Blanc. These contribute passion fruit/grapefruit/boxwood aromas, and they are also important in the character of botrytis-affected sweet wines (see chapter 17, page 196).

Finally, dimethylsulfide is a VSC that can contribute positively to the aroma of some red wines.

MOUSE TAINT

Mouse taint is a wine fault that I had not experienced until a few years ago. Now it seems to be everywhere, but that is probably because I have a strong interest in natural wine, and it is the increase in people working without added sulfites that has led to its rise. Also, there is the question of awareness. Once you have spotted a specific fault, you seem to see it more frequently. I'm going to focus on this taint because it is novel, topical, and is not covered extensively in *Flawless*.

Just how common is it? It is hard to tell. The only survey I am aware of is a graph from a webinar by Excell laboratories in France showing a graph of enquiries they have received about analysis for mouse. These rose dramatically between April 2018 and February 2019. This could reflect a rise in awareness of this fault as much as an increasing incidence. As we will see, it is a complicated fault and coming up with any sort of global figure would be hard.

In January 2019, I caught up with two of the leading French experts on mouse: Dr. Nicolas

Richard who works with Inter Rhône, and Dr. Bertrand Chatelet, who works for Sicarex in Beaujolais. We discussed the latest research on the topic. Three molecules have been identified as contributing to the taint. They are part of a group of compounds called tetrahydropyridines. They are 2-ethyltetrahydropyridine (ETHP), 2-acetyltetrahydropyridine (ATHP), and 2-acetylpyrroline (APY). These compounds are made by lactic acid bacteria (LAB) and also some strains of *Brettanomyces*. LAB and *Brettanomyces* can produce mousy taints independently. It also seems that while *Brettanomyces* can theoretically produce APY, in practice it just produces ETHP and ATHP.

Acetaldehyde (the oxidation product of alcohol) is involved in the generation of these compounds. Acylation reactions convert L-lysine and L-ornithine into the tetrahydropyridines. These compounds exist in an equilibrium in which they go from volatile to nonvolatile depending on pH. This is a critical point. At low pH they are not volatile and so you can't smell them. But when your saliva mixes with them, the pH of the saliva is higher than that of wine, and so after a while you suddenly detect them retronasally. The aromas in question are mouse cage, popcorn, sweet corn, ricecakes, crackers, grilled bread, salami, charcuterie, and dirty socks.

But, points out Richard, the perception of mouse taint is complicated. There are variations in mouth pH, variations in sensitivity, and the wine varies with aeration. "The first problem comes from the mouth pH," he says. "The variability is high. The variation for each person can be as much as one pH unit from day to day. This alters the perception of the mousy flavor." Wine pH is between 2.8 and 4, with most white wines at about pH 3.3 and most reds 3.6 on average, but this varies widely with style. Mouth pH is close to 7 and varies between 5.76 and 7.96 among individuals. With each individual, though, mouth pH will vary by an average of 0.91, depending on the time of the day, food consumed, and physiological state.

Aside from this, is there different sensitivities among people in their ability to perceive mouse compounds? "Yes," says Richard, "and there are at least three molecules. You can be anosmic for one, and very sensitive to another, and have medium sensitivity for the third. There are as many combinations as there are humans." So, the sensitivity to mouse taint varies for each of the compounds, and the sensitivity to one is independent of the sensitivity to another.

Chatelet cites an unpublished study by Université de Bordeaux researcher Sophie Tempère, who studied the ability of 23 people to detect APY, one of these compounds. She found detection thresholds that differed 1,000-fold between the least and most sensitive subjects, suggesting that some simply couldn't smell it at all. This explains the assertion that some people don't get mouse, while others do, although this is far from clear-cut because of the mass of variables involved. A figure of 30 percent of winemakers unable to detect mousiness has been cited widely, but its source is unknown. There is also a temporal effect. "The wine will vary from minute to minute," says Richard. "You can wait for thirty minutes or one hour after opening the wine, and the mousiness will appear and reveal itself." If you are opening a wine that you think may have a trace of mouse, or where you've experienced mouse in the past, serve it cold, drink it fast, and do not decant it!

As well as production by LAB and *Brettanomyces*, there is also the suggestion that the Maillard reaction might be involved in mousiness in some situations. It contributes to the flavors of bread, biscuits, and tortillas by making tetrahydropyridines, among other compounds. But this is an area of some uncertainty. Baked goods are an example of where we actually quite like the contribution of these compounds.

The risk factors for mouse are several, but the biggest is winemaking in the absence of added sulfites. High pH, wild ferments (see page 104), and potentially also the use of whole-bunch (see

page 152) are also cited by Chatelet. Richard is based in Beaujolais where he says that because of the rise of natural winemaking, mouse has become quite a problem. He thinks that part of the reason for this might be semi-carbonic maceration (see page 161). "It is a particularly high risk for mouse, because there is an interface between the liquid and solid, and there is aeration," he says. "Natural winemakers do a lot of semi-carbonic, and they have a lot of mousiness, so I made a correlation. But I have no experimental evidence."

Redox potential also makes a difference, and mouse can reappear with each exposure to air. It is quite common to open an affected bottle and for the first half to be fine, and the second half to be mousy after it has been exposed to air. Anecdotally, winemakers working naturally often say time is critical. Sometimes they have mousy wines, even bottled ones, but they just wait until they think the mouse has gone before releasing their wines.

SMOKE TAINT

This is another wine fault that is of intense current interest. Both California and Australia have had major problems with bush fires near or in wine regions that have led to grapes being tainted. As a country where this is a recurring problem, Australia has led the way in this research. The AWRI began looking at this issue in earnest in 2003 after a bad bush-fire season. Of particular interest are the following questions. What are the components in smoke that can taint wines? Which are the vulnerable periods in the vine's life cycle, where smoke exposure has the potential to cause problems? And how can you test grapes to see whether they are affected or not?

This first set of studies concluded that smoke taint was a real problem, with wines from smoke-affected vineyards displaying characters described in sensory work as "smoky, burned, ash, ashtray, salami, and smoked salmon." They also identified two compounds as key culprits: guaiacol and 4-methylguaiacol, although, probably, there were other chemicals involved, albeit at lower concentrations. They also showed that the taint compounds were found within the skins of the grapes, but not in the pulp or in the surface waxes on the skins.

Interestingly, it turns out that the chief chemical culprits in smoke taint are already present in many wines. Guaiacol and methylguiacol are formed by burning lignin, the structural component of wood. And when barrels are heated with fire to bend the staves, this results in the production of a number of flavor compounds from the wood that then find their way into wine. The concentrations typically found in barrel-fermented or aged wines seem to be fine; it is only when they are present at higher levels that they cause problems.

Further studies looked at the timing issue and found that smoke exposure at any time between *veraison* (when the berries change color and enter the final stage of ripening) and harvest can cause smoke taint, the most sensitive period being a week after *veraison* and afterward. It is also worth noting that repeated smoke exposure builds up the level of smoke taint, as might be expected. The takeaway: the less smoke the better. Initially, one thought was that it might be possible to wash the grapes while they were still on the vine to get rid of the smoke compounds, so they didn't end up in the wine. In a study from the AWRI in 2003, attempts were made to wash fruit in various ways. The treatments tested were cold water, cold water plus wetting agent, warm water, cold water plus five-per-cent ethanol, and milk. None of these worked.

This is because these flavor compounds enter the grapes pretty quickly after exposure via the waxy cuticle around the berry. They enter the berry skin cells under the cuticle, and this is where things get interesting. These cells respond to smoke compounds by a detoxification process. The technique used is called glycosylation, and it involves binding a sugar molecule to them. "Glycosylation is a defense mechanism," says Eric Hervé of California's ETS Laboratories, which has been at the forefront of looking at the science of smoke impact on grapes. "It is mostly a detoxification process: smoke compounds have some toxicity, and so the defense is to bind them to sugars and make them water soluble so they can be translocated into the cell and inter them in the vacuoles where they won't interfere with metabolism."

This creates an immediate problem. As well as making the smoke compounds less toxic, glycosylation also takes away their aromatic potential: they stop smelling. So, if winemakers want to see whether their grapes are affected, then they can't just sample them; something needs to happen to cleave the sugars off so that their aromatic potential can be revealed. And this is exactly what yeasts do during fermentation, via a process called hydrolysis. "It looks like it is yeast driven and mostly in the first days of fermentation," says Hervé. For this reason, the AWRI recommend testing smoke-affected grapes for a panel of smoke-related compounds, and then doing a sensory assessment of a small-scale ferment. This is more accurate the closer to the actual harvest date it takes place. They recommend a couple of weeks before harvest, which gives enough time for the results to be available so that the right action can be taken (see below).

What about analyzing for the glycosylated compounds? It turns out this is not so straightforward. "Our current investment in measuring these glycosylated compounds exceeds a million dollars in hardware," says Gordon Burns of ETS. "It is entirely a speculative investment so that as we go down the road we will be able to serve the industry better. We are building that very necessary background information about native levels of all these compounds." In the meantime, they believe that looking at the levels of two of the free volatiles, guaiacol and 4-methylguaiacol, is the best way to go. Even though most of these compounds will be glycosylated, looking at the small amount that is free gives a good indicator of whether the grapes are smoke impacted. "We have been doing this for ten years now," says Hervé. "Each year the fire events are different. The timing of exposure to smoke is different. The fuel is different. But over the years we haven't seen much difference in terms of the numbers. Typically, guaiacol is three to five times higher than 4-methylguaiacol. This is a good signature that it is coming from smoke."

WINEMAKING OPTIONS

So, what can be done if the grapes appear to be affected? Winemakers have a few options. The first is to try to minimize any transfer of smoke compounds from the grape skins into the must. This is easier for whites and rosés than it is for reds, which rely on extended skin contact for color and structure. Hand harvesting is a good first step, because it minimizes skin maceration. It's also important to make sure that there are no leaves in the harvest bins, because these also have smoke compounds in them. Fruit should be kept cool, because this results in less extraction, and it is ideal to press the intact whole bunches. The free run (the first 400l/ ton from the press) should be kept separate because this will have fewer smoke compounds. But this is just applicable to whites and rosés.

For reds, the received wisdom is that shorter extraction time is better than longer, but ETS dispute this. "Smoke compounds are mainly in skins so if you prevent skin contact you minimize the problem," says Hervé. "This works beautifully for whites or rosés, but for reds this doesn't work in our experience. The smoke compounds migrate into fermenting juice extremely quickly." They

have studied this. "In all the red ferments we have followed, the maximum number of smoke markers are already extracted after three days. Trying to shorten the fermentation with reds is actually making the situation worse because it is a lighter wine that shows up smoke impact more clearly."

There are some other winemaking techniques that can be used to mitigate smoke impact, but they all have their downsides. One is to fine with a product called activated carbon, which is already used in some winemaking settings. Carbon grabs hold of organic materials and removes them: for this reason, it is widely used during filtration in water treatment. The problem is, it is not particularly selective. It can be used for treating juice or wine, but it has a negative effect on other aroma compounds. Another approach has been to use reverse osmosis, which is a specialized filtration process, coupled with a technique called solid-phase adsorption. Reverse osmosis takes out a permeate fraction from the wine that contains water, acids, alcohols, and chemicals below a certain molecular point cut-off. This fraction is then treated alone with solid-phase resorption to take out the smoke compounds before returning it to the wine. It removes some beneficial aroma compounds, too, but the hope is that because just this small fraction of the wine is treated, its negative impacts will be limited. One of the problems with this approach is that while treatment has been effective in some cases, the smoke impact came back after a while, presumably because of the ongoing presents of some glycosidases that later hydrolyzed to reveal smoke compounds that were not removed, so wines treated this way have to be drunk quickly.

15 The evolution of élevage: oak, concrete, and clay

Élevage is a French term used to describe the way a wine develops in the cellar. The literal translation is "breeding," but I like the suggestion that it involves bringing-up the wine, from its infant state post-fermentation to its first maturity, when it is bottled. There is great skill in good *élevage*, and with all the talk of wine being made in the vineyard, the appreciation of this art is somewhat lacking. One of the big changes occurring in the world of wine—particularly at the top end—is a revolution in the cellar. Go back 100 years and wine cellars would have been full of large oak vats, big barrels, and concrete tanks, as well as smaller oak barrels. Modernity brought stainless steel and, for certain wine styles, an obsession with new small oak barrels. Now many are returning to the past: large-format oak, concrete, and even terracotta. In some ways, it is a recovery of a lost art, while retaining the benefits of a modern understanding of wine science. This chapter looks at the science of *élevage* and the different ways it is practiced.

BARRELS

Barrels are probably the earliest form of wine technology, and their use is still vital in the production of many of today's wine styles. Despite their importance in the winemaking process, the positive flavor effects of barrels were likely discovered providentially. They just happened to be the best way of storing and transporting liquids, and until the advent of cement and stainless-steel tanks, winemakers lacked alternatives. However, oak's accidental association with wine has been a critical one. Many fine red wines are dependent on oak barrels for a vital component of their flavor, as are a good number of whites. Without oak, wine would be very different altogether. Even where older, larger barrels—which don't have such a direct impact on flavor— are used, their ability to allow exposure of the contents to small amounts of oxygen is important in the development of the wine.

Above Barrel construction. The staves are assembled and kept in shape by rings, and then the barrel is toasted over a brazier to heat the staves in order to allow them to be bent. The toast also helps the barrel transmit flavor to the wine. This is at Tonnellerie Mercurey in Bourgogne, France.

OAK

Let us begin with a slightly tangential biological perspective. Taking a somewhat simplistic conceptual view, there are four organisms crucial to wine production: two microbes and two woody plants. The microbes are the yeast *Saccharomyces* and the lactic acid bacterium *Oenococcus*. The plants are the grapevine *Vitis vinifera* and the oak tree *Quercus*. Of these, two are essential for all wine styles (grapes and yeast), but red wines and some whites need lactic acid bacteria, and many red and white styles would not be possible without oak.

The genus *Quercus* can be split into many hundreds of species. There are four that are principally relevant to wine, three of which are used to make barrels (*Quercus alba*, *Q. sessiflora*, and *Q. robur*) and one of which makes corks (*Q. suber*); *Q. robur* also goes under the name *Q. pedunculata*, while *Q. petraea* is a synonym for *Q. sessiflora*. Taxonomy is a confused branch of science.

Why is oak so good for barrel construction? It is strong yet still relatively easy to work. It also has the capacity to make containers that are watertight. Its wood is rich in structures known as tyloses. Wood is mostly composed of large, open, water-conducting vessels, the xylem vessels, which travel up the trunk. The tyloses are outgrowths from neighboring cells that block the xylem vessels. American oak is particularly rich in these structures, and thus the wood can be sawn in a number of planes and still be impermeable, while French oak has fewer and has to be split in specific planes for it to make watertight staves. But perhaps most significantly, oak facilitates and is also directly involved in chemical interactions with wine that have positive effects on its flavor and structure. That is why, in this technological age, barrels still have not been replaced in the cellar.

BARRELS: THEIR FLAVOR IMPACT ON WINE

I remember the first time I sampled through the barrels in a producer's cellar. I was struck by the differences among samples of the same lot of wine that differed only in the barrel they were being aged in. These barrels varied in their toast level, manufacturer, and source of oak. For those who use them, the choice of barrel is an important winemaking decision, and skilled producers will be as fussy about the barrels they use as they are about the condition of their grapes.

There are several factors that influence the way that barrels affect wine flavor. Typically, oaks used in barrel production are classified on the basis of their geographic origin. The first, and most important distinction is between French and American oak. American oak is a separate species (*Q. alba*) with quite different characteristics from the two French species (*Q. robur* and *Q. sessiflora*). Within the category of French oak further subdivisions are made according to the forest region, which closely (but not completely) correlates with species used.

The situation is further complicated by the fact that each cooper has their own unique house style, and the characteristics of the staves will differ according to factors such as age of tree, the part of the trunk they were taken from, the seasoning process, and the amount of toast. This interplay between oak species, environment, and human intervention makes the science of barrels almost as complex as that of the viticultural and winemaking processes.

Above Different toast levels in oak. This affects the flavor impact of the barrel, and (along with the choice of the type of oak) is an important decision for winemakers who choose to use new barrels.

Different species of oak used in barrel production[a]

Name	Origin[b]	General characteristics
Quercus robur, known as pedunculate oaks	French forests, principally Limousin, Burgundy, and south of France	High extractable polyphenol content; makes wines that are more structured, less aromatic
Quercus sessilis, known as sessile oaks	French forests, principally in the Central and Vosges regions	Contributes more aroma and less structure
Quercus alba	America	Low phenol content, very high concentration of aromatic substances[c]

[a] Because of the high cost of French oak, central European alternatives are now being considered. Tonnellerie Lafitte is a sister company of Vicard that specializes in Hungarian oak.

[b] The three most commonly encountered French barrel styles are Nevers, Allier, and Tronçais. These encompass the wood from these areas but are also terms used to "type" wood from other regions.

[c] American oak has extremely high concentrations of oak lactones. For example, one study by Pascal Chatonnet showed that French sessile oak gave concentrations of methyl octalactone of 77µg/l, French pedunculate 16µg/l, while American oak delivered a whopping 158µg/l.

Coopers are interested in the central part of the trunk, the dead, tough heartwood that is also known as "stave wood." This is split along medullary rays (horizontal structures that run radially through the wood). Splitting is essential for French oak because if it were sawn, it would be porous. American oak can be sawn because of the presence of tyloses. The splitting continues until single staves are produced.

The fact that French oak must be split while American oak can be sawn in part explains why French oak barrels are more expensive than their American oak counterparts. The house style of various coopers is probably the most significant factor influencing the effect of the barrel on the wine.

Before oak is used for barrel construction it must be seasoned. This is in order to bring its humidity levels into line with the environment it will be used in, and to allow some important chemical modifications to occur. This typically takes two or three years, depending on the thickness of the staves. Seasoning is a bit of a balancing act. You want to leave the wood long enough, but not too long. Seasoning normally takes place outdoors, and results in a number of changes to the wood: ellagitannins are reduced, as are levels of bitter-tasting compounds called coumarins. At the same time, there is an increase in some aromatic components, such as eugenol. It is possible—and less expensive and quicker—to age staves in ovens, but the drawback is that these important chemical changes do not occur. The oak has fewer aromatic properties and more bitter compounds ready to be leached into the wine.

The barrel-manufacturing process involves heating the staves over a brazier so that they can be bent into shape. Somewhat fortuitously, this slight charring—or, "toasting" as mentioned here already—coupled with the chemical properties of the wood, means that the interaction of the wine with the inside of a new barrel imparts pronounced flavor characteristics to the wine. When used appropriately, new barrels can have a significant beneficial impact on the wine that is aged in them. These are summarized in the table. There are variations on the toasting process, which can result in different levels of char.

Flavors from oak[a]

Flavor compound	Characteristics	Influence of the manufacturing process
Lactones	The most important oak-derived flavors in wine are cis and trans isomers of β-methyl-γ-octalactone, known as the oak lactones. On their own, these oak lactones smell like coconut, but in wine they can smell quite oaky, too. The cis isomer is described as having an earthy, herbaceous character, as well as the coconut, while the trans adds spice to its coconut aromas.	Seasoning barrels affects the ratio of cis to trans forms of oak lactone, and toasting is thought to reduce overall lactone levels. American oak contains much higher lactone concentrations.
Vanillin	The main aroma component of natural vanilla. This is present in significant quantities from oak wood and contributes significantly to the aroma of oaked wines. If the wine is actually fermented in oak barrels, yeast metabolism reduces the vanillin concentration by reducing it into the odorless vanillic alcohol. Therefore, barrel-fermented wines smell less oaky than the same wines would that were fermented in a tank and then transferred to a barrel, even though they have been in oak for longer.	Levels can be increased by toasting but decrease at high toast levels.
Guaiacol	Guaiacolol and the related 4-methylguaiacol have a charlike smoky aroma. 4-methylguiacol is also described as spicy.	Formed by the degradation of the wood component lignin during toasting and are therefore increased at high toasting levels.
Eugenol .	With a clovelike smell, this is the main volatile phenol associated with wood. The related isoeugenol smells similar.	Increased during the seasoning process and reported to increase during toasting.
Furfural, 5-methylfurfural	Produced by the heat-induced degradation of sugars and carbohydrates, they have sweet butterscotch and caramel aromas, with a hint of almond.	Made when carbohydrates in the wood are degraded by the heat of the toasting process.
Ellagitannins	Tannins absorbed by the wine from the wood are known as ellagitannins. They are capable of modifying the structure of the wine, as well as combining with anthocyanins and increasing color. They have an astringent taste. Ellagitannins belong to a class known as hydrolyzable tannins, and the two main isomers are vescalagin and castalagin.	Their concentration decreases at heavy toasting levels.
Coumarins	Derivatives of cinnamic acids that are present in oaked wine at low concentrations but which still affect flavor: their glycosides are bitter and their aglycones acidic.	

[a] Many of these compounds occur at levels below their individual detection thresholds in wine. Yet they can still have an effect on wine flavor and aroma through important synergistic effects. For example, the perception threshold for oak lactone is reported to be 50 times lower in the presence of vanillin. In addition, the combination of more than one of these can produce complex flavor/aroma sensations. Whether or not any of these flavors will be positive depends on the context of the wine. It is a complicated business.

MICROOXYGENATION— THE TRADITIONAL WAY

Barrels do not only impart flavor directly. Another equally important, but less talked-about effect of aging wine in barrels is the way it allows a very slight and controlled exposure to oxygen. Normally, winemakers do all they can to avoid exposing their wines to air, but in this case the low-level oxidation that barrels permit is beneficial to the structure and character of many wines.

Wines stored in average-sized barrels (around 225l/60 gallons) typically receive between about 20 and 40mg of dissolved oxygen/liter/year, but this is difficult to measure precisely because some is consumed by ellagitannins in the oak. Some oxygen passes through the wood itself, the majority passes through gaps between the staves, and the remainder comes through the bunghole.

This low-level exposure to oxygen has a number of important effects. Color is intensified because of reactions between tannins and anthocyanins, and tannins are typically softened by polymerization, which eventually causes them to precipitate out of the wine. But in the shorter term, barrels can help build a wine's structure in much the same way the microoxygenation does.

Some companies, such as California's ETS Laboratories, are now offering analytical techniques that will allow winemakers to test the performance of barrels by analyzing their chemical imprint. To explain the results in ways that winemakers can visualize, they use a graphical representation known as a radar plot. Dr. Eric Hervé of ETS gave me an example to explain how they do this. "Imagine that a winemaker wants a better understanding of how barrels from various origins influence the aroma of one of his top-end wines. The trial is simple: a same lot of wine is aged in groups of barrels from various coopers, with various specifications (i.e. toast level). During the aging, the wine is tasted and analyzed. We report concentrations of oak aroma compounds, for each sample, in micrograms per liter [μg/l]. Winemakers can then build a database for future reference and comparison. We also generate a 'radar plot,' for each sample, by comparing concentrations found versus the 'trial average' [mathematical average of concentrations found in all samples from the trial]. This allows us to see immediately what aroma compounds are below or above average, and therefore if associated aroma descriptors are likely to be less or more intense, compared to the other wines in this trial." In this way it is possible to compare barrels from different coopers, different species of oak, and even with different toast levels. A strong relationship of trust has to exist between winemakers and their barrel suppliers, because once wine has gone into a bad or unsuitable barrel, it is too late to reverse the decision. But this sort of analytic technique is likely only to describe a small fraction of oak-imparted-wine-flavor compounds. Do these sorts of analyses measure all the significant molecular contributions of barrels, or just a few? I asked Hervé about this. "Although the compounds we measure are widely considered the main contributors to the oak flavor, there are many others that may play a role in wine," he responded. "Luckily, most of these compounds belong to a 'family' of related compounds [same origin and similar odors] and measuring one or two molecules only gives a good idea of the 'family' as a whole. This is the concept of 'indicator compounds.' One good example is the volatile phenols family: degradation of oak lignin by heat gives hundreds of these molecules, several of them playing a sensory role in wine, in synergy. Measuring only the two most important ones (guaiacol and 4-methylguaiacol), however, proves to be a very reliable indicator of the 'smoky' character in wines."

OLDER BARRELS

Not everyone likes or wants the flavor imprint that new oak barrels stamp on a wine. While certain styles of wine have got the substance to absorb flavor compounds from new oak without being dominated by them, many wines are best aged in second-, third-, or even fourth-use barrels. Older barrels impart progressively fewer flavors to wine, but they still allow the controlled oxygen exposure that is important for resolving structural elements in the wine. While it is possible (and important) to clean barrels between uses, it is impossible to sterilize them. If they harbor spoilage organisms such as *Brettanomyces*, there is virtually no way to remove the potential inoculum because of the porous nature of the wood. There is always somewhere for the bugs to hide. Such barrels can still be used, but they need to be monitored very carefully.

ALTERNATIVES TO BARRELS

Barrels are expensive. Oak chips, staves, and even liquid-oak extract have been used to give cheaper wines some oak complexity and flavor. The results are mixed, and rarely replicate the characteristic imprint of barrels. However, the use of oak alternatives suspended in tanks during fermentation of red wines offers a means of adding some barrel fermentation character to red wines. Usually, barrel fermentation is not possible for red wines because of the fact that they are fermented on their skins, and oak interaction is usually restricted to the post-alcoholic fermentation period. This may offer some interesting options for experimentation by winemakers. It could increase structure development and enhance color stability.

HAS THE WINE WORLD FOCUSED TOO MUCH ON SMALL OAK?

Over recent years there seems to have been a move away from small oak barrels (of the 225- and 228-liter [about 60–gallon] capacity common in Bordeaux and Burgundy) toward larger barrels and even the elimination of barrels altogether in *élevage*. It has been suggested that the increase in demand for oak has led to some quality issues in barrel production. More significantly, many winegrowers have decided that they don't like the flavor of new oak in their wine, and that the oxygen exposure offered by smaller oak may not suit all wine styles. Chris Alheit, a winegrower in South Africa, likens new oak flavors in wine to retsina, where pine resin is added to wine.

Another South African winegrower, Eben Sadie, has been moving away from new small wood to bigger-format containers, such as big old oval casks and concrete. "Cabernet family grapes do really well in small barrels," Sadie maintains. "All varieties with certain tannins need a rapid evolution in order to be bottled at 18 months, but it has become clear that Mediterranean varieties are completely different. They already have advanced tannins and more fragile fruit."

This move away from small oak has proved to be a massive logistics exercise. Sadie is getting big oak from Austria, and buys one or two large barrels a year, as well as using concrete and terracotta. "I have begun to dislike wood. But it is because of where I am," he says. "People making wine in continental climates can really use wood and complement their wines. But for us, working with very mature fruit, it is more difficult. The barrel is essentially an incubator for ripening the wine. You ripen the wine in the wood. For us, with a Mediterranean climate, where the grapes are inevitably mature, the last thing we need to do is mature the wine more in the wood. The tannins are mature. What we have to do is move away from oxygen and protect the fruit and freshness of the wine. So, I am moving away from wood. Also, on a generic profile across the world, people are moving away from wood. I think it will be an interesting move because we will taste more wine. If you look at a vineyard that is one hundred years old, why in the world would you put these grapes into one hundred percent new wood and make this wine taste like a tree that grows in France for at least ten years of its life? Eighty percent of consumers will taste the tree in France and not your terroir, because most people drink the wines before they are ten years old."

Above Made in clay: *talha* is the name for the large amphorae used in the Alentejo, Portugal. This is a collection of talha at Herdade do Rocim, a winery that hosts the annual amphora wine festival.

Above Amphorae from Artenova, in Italy, being used in a cellar in Washington State. Even big wineries often have an amphora or two lying around, possibly because winemakers love experimenting.

Above An amphora farm at Brash Higgins in the McLaren Vale, Australia. The use of terracotta vessels for fermentation or aging is becoming much more common, even though these are expensive, fragile, and hard to clean.

Sadie's cellar does have some oak barrels in it, though. "We only have one wine that we still use small barrels for, and that's Columella," he says. "I tried to make this wine outside new oak, and it doesn't want to be geek or hipster, it just wants to be wine." But whereas he used to use 60-percent new oak, the barrel program has changed. "We only buy four new barrels a year, and we throw out four, and there are forty barrels in here," says Sadie, speaking in his Columella cellar. Some barrels are ten years old.

CLAY

There has been a revival of interest in using clay for *élevage*. Such is the growing use of amphorae, *talha, tinajas,* and *qvevri* that there is even an annual festival, held in the Alentejo in southern Portugal, focusing on the use of these vessels. They have a tradition in this region of using large amphorae called *talha* for making wine in, something that was dying out, but has recently seen a revival.

I attended the second edition of this *talha* festival, held at the Herdade do Rocim winery, in November 2019. David Rego, export director of the estate, explained to me that in 2018 they had 1,200 attendees and were bursting at the seams. They restricted tickets to the same number in 2019, and it was a sell-out.

Rocim started bottling its *talha* wines seven years ago and Rego says that at the time everyone laughed at them. They have 26 *talhas* now and would like more. "It is tough to find them," says Rego. "You really need to dig around to find someone who has a *talha* at their place that they aren't using." Rocim don't use sulfur dioxide during *talha* vinification, but do add some before bottling. "For those who don't bottle, sulfur dioxide is not on the menu," says Rego.

The typical way of using *talha* in the Alentejo is to use some whole-bunch, or even completely whole-bunch fermentation. The amphorae are lined with *pez*, which is a mixture of pine resin and beeswax. The top of the amphora is not sealed. Instead, there is a layer of olive oil on top of the wine, and a cloth will be placed over the opening to keep bugs out.

QVEVRI

Georgia is the country most associated with fermentation in clay. On my first trip to Georgia, in 2019, I went to the village of Vardisubani in the Kakheti wine region, to meet a fourth-generation qvevri-maker, Zaza Kbilashvili. In the basement of his house there are several large clay vessels in the process of being built. They are already almost as tall as me, but there is clearly quite a bit yet to be added to each. It is quite cramped and dark in the basement, but the smell of damp clay is intoxicating—it takes me back to pottery classes at school, among some of my favorite lessons. Qvevris (sometimes spelt kvevri) are the traditional vessel for making wine in Georgia and are, to the outside world, the symbol of Georgian wine. But unlike in other countries where terracotta amphora are freestanding, these qvevri are buried in the earth. When you enter a traditional Georgian wine cellar—a marani—you only see the qvevri openings. Marani floors are usually made of bricks, and every now and then there is a brick circle, surrounding the qvevri mouth.

Georgia's history of wine production is thought to go back 8,000 years. This unique way of making wine has survived partly because this heritage of winemaking has followed an unbroken lineage. In the past there were many villages that specialized in qvevri manufacture. Now it is restricted to just a few, in three regions: Kakheti, Imereti, and Guria. Kbilashvili's profession was dying out, but with the recent attention that qvevri wines have achieved, his order book is now full. The demand is not just local, as there are also plenty of orders from abroad. But should you want to buy one, you will have to wait two years. The cost? They are surprisingly affordable, at 2 Georgian Lira/liter. This means a 3,000l (790-gallon) version costs 6,000 Lira ($2,100). If they are broken, they cannot

be repaired, but well maintained they can last for centuries. His facility produces around 50 a year.

Qvevri of different sizes exist. They can be 6,000–8,000l (1,585–2,110 gallons), but most are 2,000–3,000l (530–790 gallons). One of the big advantages of qvevri is that, being sunk into the ground, they maintain a fairly steady temperature. The other advantage is that this way of making wine means that, if a good degree of hygiene is maintained, no additives are needed. The qvevri has become a champion of the natural wine movement, although not all qvevri wines are natural wines.

The shape of the qvevri is thought to be important. The bottom is like a pointed cone, and at the end of fermentation, the seeds separate from the skins and fall to the bottom of the vessel. These are then covered by the lees and are kept away from the wine. So, the wine finishes fermentation and ages in contact only with the skins and lees and doesn't extract the more bitter seed tannins. The tannins present in the wine bind up any proteins that would otherwise stop the wine from appearing bright and clear, and they clarify naturally in the qvevri. In the past, there is evidence that winemakers used to add crushed flint to the wines to help with removal of tartrates.

How qvevri are made

Kbilashvili explains how the qvevri are made. The clay comes from the local forest, and after it is cleaned to remove stones and debris, it is ready to go. He shows us how they start, with the base. Then, after this, the walls are built up a little by little, around 20cm (8in) at a time. How often this is possible depends on how long it takes the previous layer to dry enough to apply the next, which is weather dependent. This construction is all done by hand and eye: no template is used. The maximum capacity he can produce is 3,000l (790 gallons).

Once the qvevri is finished—around eight are built at any one time, because this is the capacity of the kiln—it is allowed to dry. Then

Kbilashvili enlists friends to help take the large vessels carefully to the bottom of the garden to the brick-built kiln. It takes six people to move one. Once they are in place, the door is bricked up, and the long process of firing begins. The fire must reach temperatures of 1,200–1,300°C (2,192–2,372°F). The hotter the fire, the less porous the *qvevri*. This temperature has to be maintained for several days, so they take shifts monitoring the fire. It's quite a process.

It is possible to make wine in unlined *qvevris*, but mostly they are waxed, especially when the pore size is larger due to a lower firing temperature. Usually, the *qvevri* is waxed before it is put into the ground, but it is possible to do this later when it is already buried. The *qvevri* is heated by lighting a fire inside, using dried vine cuttings, and turning it slowly. Once it is hot, the ash is removed. Melted beeswax is added and is soaked up by the *qvevri* walls. The wax is applied using a piece of cloth on the end of a long stick, working from the bottom up. Buried *qvevris* are waxed by lowering an open-topped metal cylinder with a fire burning in it into the vessel (the chimney has to be long enough to take the smoke out of the *qvevri*). It is important that the wax layer is not too thick since the wine needs to be in contact with the clay for the proper *qvevri* experience.

The outside of the *qvevri* is usually coated with lime. Sometimes they are coated with cement, but lime is better because it lasts much longer. Liming improves strength of the *qvevri* and helps with the winemaking process. Especially when the *qvevri* is buried in damp soil, lime is much better than cement. The liming is traditionally done with a lime grout that is 1kg (2.2lb) lime to 2kg (4.4lb) of sand, together with rubble and sandstone fragments and perhaps fragments of old *qvevri*. Typically, this would be done in situ, as the *qvevri* are being placed in the ground in the new *marani*. The thickness of this lime is 10–25cm (4–10in).

Washing is a critical phase. In the past, this was done by spraying lime water on the inside of the *qvevri*, pouring it out, and replacing it with boiling water. The lid was then replaced and then a while later the softened dried pomace and lees were removed. To prepare the lime wash, the lime is burned and then dissolved in the water. The water is separated from the precipitated lime and then used to wash the *qvevri*. The wall is cleaned by using a special brush formed from the roots of St. John's Wort, or the bark of cherry trees. Sometimes ash water can be used instead of lime water. To make this, wood ash is boiled (1.5kg/3.3lb

Above The *qvevri* is the traditional fermentation vessel in Georgia. These large clay vessels are coated in lime or cement and then set into the ground, where they stay. A winery with *qvevri* is known as a *marani*.

Above Saperavi fermenting in a *qvevri*. Both white and reds are fermented in these vessels. Saperavi is Georgia's leading red variety, and is a *teinturier* (red-fleshed) grape capable of making deeply colored wines.

per 3–5l/0.7–1.1 gallons of water) and the wash water is drained off the sediment. Sometimes raw ash is applied to the inside of the *qvevri* while the walls are wet. Here it dries, protecting the vessel if it is to be left empty for a while.

Sulfur wicks are often burned inside the *qvevri* to sterilize them, typically 3g (0.1oz) per 100l (26 gallons) of volume. It is best if this is done while the walls are still wet.

A *qvevri* lid is usually made of wood or stone. Stone lids were very popular in eastern Georgia and were made from slate from the slopes of the Caucasus. The wooden lids are made from linden, chestnut, or oak. The tradition in the east is to use wet clay on the surface of the *qvevri* neck, and then add the stone lid. In the west, they added the wooden lid and then covered this with earth.

In the east, in Kakheti, it was traditional to use all the stems and skins in the ferment. Sometimes the grapes would be crushed a bit first, but the entire skins and stems would be placed in the *qvevri*. For whites, the skin contact would extend for six months. For reds, it is shorter: just the duration of the fermentation, and then the wine is pressed back to the *qvevri*. In the west, in Imereti, it was normal to use only around 30 percent stems. There are lots of variations of winemaking in *qvevri*, we found, as we visited a range of different *marani* in the region. The interesting thing is that usually the whites have longer skin maceration than the reds.

CLAY AROUND THE WORLD

Eben Sadie, Duncan Savage, Donovan Rall, Catherine Marshall, and Hamilton Russell are some of the South African producers who are using amphora made locally by a potter called Yogi de Beer. "Surely there is more to winemaking than stainless steel and oak," says Duncan Savage. He first contacted de Beer several years ago when he was winemaker at Cape Point Vineyards, after he had become interested in using clay amphorae as tools for *élevage*. Savage didn't want to import them, so he worked with de Beer,

Above *Qvevri*-maker Zaza Kbilashvili standing in front of the building where the *qvevri* are fired. There are relatively few *qvevri*-makers left, but demand for these vessels, which are made by hand from clay found in the woods, is on the increase.

who is now throwing 600l (about 160 gallons) pots for winemakers to use. Savage describes this experimentation as a "cool journey." The first pots he tried were 120l (about 32 gallons) because the potter he used had only a small kiln. They were earthenware and had been fired at low temperatures, so when Savage poured wine in, they leaked like sieves. The ones he uses now are stoneware clay made at high firing temperatures and are unlined. De Beer's largest amphorae are 860l (about 230 gallons) in capacity.

Savage is following in the footsteps of winemakers like Josko Gravner from Friuli and Frank Cornelissen and Giusto Occhipinti (of Cos from Sicily), all of whom use amphorae in the upbringing of some of their wines. Gravner's evolution as a winemaker took him from stainless steel, to large format oak, and then—after a trip to Georgia in 2000—to *qvevri* sealed with beeswax and sunk in the ground for his whites (in 2001) and later for his reds (in 2006). His view is that you aren't really using amphorae (which are imported from Georgia, one of the few remaining countries with the facilities for firing large examples, as explained in these pages) if you only have a few.

Now, every wine he makes is fermented in

Above In Bourgogne, Michel Magnien have switched a large portion of their production to these terracotta amphorae. The results are impressive.

Above Clay amphorae made by Cape Town potter Yogi de Beer, in the cellar of Sadie Family Wines, Swartland, South Africa. The firing temperature determines how much oxygen is transmitted to the wine.

them, and he has 45 of them in the cellar. "You are not able to ride two horses at a time. If amphora is the best way to make wine, then all the wine has to be made like that. I am not against industry or technology," says Gravner, "but we as bodies are the same as we were two thousand years ago."

In Oregon, Andrew Beckham is a ceramics teacher in the Willamette Valley who also makes some very interesting wines with his wife Annedria. Their A.D. Beckham wines are matured in terracotta amphorae that he has made, and now that he has purchased a commercial rig, he has begun commercializing his terracotta amphorae, *tinajas*, and *qvevri*. They are proving very popular.

In the McLaren Vale, Australia, is a producer called Brad Hickey whose winery, Brash Higgins, has an amphora farm. He has quite an array of these clay fermenting and maturing vessels, and they are producing some interesting wines.

Champagne is thought of as a very traditional wine region but, even here, there is experimentation underway with alternative *élevage*, and not just stainless-steel tanks and small oak. Rodolphe Peters of famed grower Champagne Pierre Péters uses a mixture of stainless steel, concrete egg, and large oak for his reserve wines,

and all taste quite different, with the concrete being the most interesting. Anselme Selosse is now using some terracotta amphorae in his cellar for fermenting base wines. And Benoit Tarlant has some base wines that are kept in cylindrical ceramic spheres called Clayvers.

In the UK, Ben Walgate at Tillingham Wines in Sussex has a number of *qvevri* buried in the floor—an English *marani*—but also a couple that are supported in sand in macrobins on a pallet, so they can be moved around.

CONCRETE

Concrete tanks used to be common in wineries and did a very good job. It has good thermal inertia, meaning that the ferment does not get any thermal shocks along the way. The tanks were often replaced with stainless steel as wineries modernized. Now people are moving back to them. A new addition on the scene has been concrete eggs, which are now very popular, and also tulip-shaped concrete tanks for fermentation. Well-known manufacturers are Nico Velo in Italy, Nomblot in France, and Sonoma Cast Stone in the USA. Concrete can be raw or can be epoxy-lined. If it is raw, it is commonly painted with tartaric acid

Above Clay and concrete dominate the cellar at Okanagan Crushpad in Canada's British Colombia. This pioneering winery is making some of Canada's most interesting wines.

Above Concrete tanks of varying sizes and shapes at Carmen winery, in Chile. A decade or so ago, a large commercial cellar like this would have been dominated by stainless-steel tanks and oak barrels.

before wine is put in. Advocates of concrete eggs say that their shape keeps yeasts in suspension longer, adding to the texture of the wine.

Spanish winery Ramón Bilbao researched different types of concrete when planning the new cellar for its high-end Rioja project LaLomba, in conjunction with researchers at Universidad de Vallodolid. They obtained blocks of concrete of the same thickness from different companies and sent them to the scientists to investigate their oxygen transmission rates. They compared concrete with epoxy covering to raw concrete and concrete with a tartaric acid covering. Raw concrete gave oxygen transmission similar to a barrel while epoxy-covered concrete gave no oxygen transmission. Tartaric reduced oxygen transmission by 50 percent.

CONCLUSIONS

Across the wine world, one of the biggest changes in the last decade has been the attitude of winemakers to alternative *élevage*. Even in large commercial wineries, it is quite normal to see concrete eggs, or some terracotta, and manufacturers of large-format oak, such as Austria's Stockinger, are doing a roaring trade. The science behind these alternatives to stainless steel and small oak is still somewhat anecdotal, but it is great to see so much experimentation.

16 Flotation and stabulation

In this chapter, I will take a look at two increasingly widely adopted winemaking techniques that have some interesting science behind them: flotation and stabulation.

MUST CLARIFICATION BY FLOTATION

How clean do you want your juice? When white wines are pressed, the grape juice that comes out is pretty murky. The grapes are sometimes covered in dust, but even when they are clean, the juice is not clear. (As an aside, there is a winery in Franciacorta, Ca' del Bosco, that washes its grapes before fermentation in a "berry spa."[1]) There are some winemakers that take the juice straight out of the press pan to barrel, but this is quite rare, and can result in high-solids fermentation in some of the barrels that then risk the development of reductive aromas. It follows that winemakers looking for a bit of reduction sometimes deliberately do a high-solids ferment. I also know some winemakers who take the juice from the press, then stir it in a tank before going to barrel so that each has the same level of high solids: here, they are deliberately toying with a bit of reduction to get some matchstick/flint character into the wine. But, in most cases, after pressing whites and rosés, it's normal for winemakers to clarify the must before fermentation.

The traditional way of doing clarification is cold settling. Here, juice goes from the press to a tank and is usually kept overnight at a low-ish temperature (to stop fermentation beginning) in order to allow the solids to fall to the bottom. The next day, or when the juice has cleared to the chosen level, it will be decanted off and fermentation is started. Sometimes pectolytic enzymes are used to help speed the process. Winemakers often have an idea of the sort of clarity they want in the juice. It is measured in terms of NTUs (nephelometric turbidity units), and from the press the typical level would be 500–600. For most winemakers, around 100 is

---

1 What does washing grapes achieve? There are no peer-reviewed scientific papers on it, but I did find a paper from an Italian group led by Agostino Cavazza presented at a symposium on Microbial Food Safety of Wine in Vilfranca de Penedès, Spain in 2007. The authors evaluated the effects of grape washing at the Pojer e Sandri winery in Trentino, in a winery-scale trial during the 2007 vintage. They used 10,866kg (12 tons) of Müller-Thurgau and 3,629kg (4 tons) of Cabernet Sauvignon, washing half of each. Then they compared three treatments for each: spontaneous ferment, inoculated ferment, and ferment started with a *pied de cuve* (a small, prior spontaneous fermentation that is then checked for quality and, if good enough, added to the must to get things going). The washing unit was the CLU Grapeclean unit from Technicapompe Zanin. The grapes were washed for three–six minutes in a one-percent citric-acid solution bubbled by air, and were then rinsed, and dried in an air flow. The results showed a difference. The must of the unwashed grapes had a higher yeast content, mostly non-*Saccharomyces* species. But the wild ferments worked fine in both cases. The washing also removed traces of pesticide, although these were low in the unwashed grapes because it had been a dry season. For the red wines, there were lower metal-ion levels (iron, copper, zinc, and lead) in the washed grapes. There was no sensory work on the wines looking at the flavor effects of washing.

the optimum level, but some go as low as 50. The benefit of working with cleaner juice is that often the wines are cleaner and fruitier. But if the juice is too clear, then there can be problems with fermentation as there might not be enough nutrients to feed the yeasts.

Flotation is an alternative way of clarifying must. It is not a new technique, but was transplanted to wine from the Australian mining industry in the late 1970s. It only really took off fairly recently, and is now becoming widely adopted. A survey of Australian wineries in 2016 indicated that they considered it to be the second most significant recent technological change in their winemaking after the cross-flow filter.

The concept behind flotation is quite intriguing. Rather than let the solids slowly settle to the bottom of a tank, nitrogen or air is introduced to the must at pressure. Then small gas bubbles form in the saturated juice and as they float to the top they take the solids with them (these bubbles combines with solids are called "flocs"). The cleared juice is then taken from below this foam.

It is common to add things to the must to facilitate the process. Pectolytic enzymes are often used to break up compounds from cell walls called pectins, which cause the juice to be hazy and more viscous. The pectins have a negative charge and surround positively charged proteins in the juice. These enzymes aid clarification by breaking the pectins into smaller fragments. The other addition is adjuvants such as gelatin, bentonite (a montmorillonite clay that is also used to fine white wines before bottling to make them heat stable), and silica. Bentonite is negatively charged and attracts positively charged proteins. Gelatin is mainly negatively charged and strengthens the flocs, helping the gas bubbles to stick to them. Silica helps create large flocs, and can be used with gelatin, but needs to be added first. These adjuvants inactivate enzymes, so if oenological enzymes are used they must be

added first. There are also some plant-protein adjuvants that can be used instead of gelatin (which is animal-derived) for vegan-friendly wines.

BATCH OR CONTINUOUS FLOTATION

There are two ways of doing flotation: batch or continuous. When flotation is done in a single tank using an adapted pump, the process is known as batch. The adjuvant (if used) is added to the tank, and then the must is saturated with the gas (to a pressure of five–seven bars). Flotation takes place and the clarified liquid is transferred to another tank.

Continuous flotation involves specialist tanks and a more complicated setup, but it is faster. Here, there is a separation tank where the must is treated with adjuvant and then saturated with nitrogen or air. It is then transferred to a shallow, cylindrical flotation tank where a rotating arm with a suction device scoops the floated foam off of the top, while the clarified juice is continuously removed from below. The floats that have been sucked off the surface can then be filtered with a lees-recovery filter (known as RVD for "rotary vacuum drum"), a centrifuge, or a high-solids cross-flow filter. Batch flotation can typically do 5,000l (1,320 gallons) per hour, whereas continuous flotation can do 40,000l (10,566 gallons).

One choice facing winemakers who use flotation is which gas to choose. Argon is inert and would work well but is too expensive. Nitrogen, which is also inert, is usually the gas of choice. Compressed air is the cheapest and is sometimes used. But air contains 20 percent oxygen, so for some wine styles it is not appropriate as it might oxidize some juice aromatic compounds. But here we hit on one of the choices winemakers face in handling white musts. While protecting juice aromatics is good for some grape varieties and styles (such as Sauvignon Blanc), in other cases it is actually beneficial to expose must to oxygen, because this can oxidize some of the juice phenolics that can later cause

Above Continuous flotation setup. The wine is saturated with gas, and an adjuvant may be added. Then, in the flotation tank, solids float to the top and are skimmed off.

problems for white wines, making them more prone to oxidation. This benefit outweighs the loss of the juice aromatics and is commonly used with Chardonnay musts. For flotation, carbon dioxide does not work well because the bubbles are too big.

The great advantage of flotation is that it saves time. Juice can be in the fermentation tank within hours rather than a day. This technique has been particularly well received in New Zealand's Marlborough region, where Sauvignon Blanc is the dominant grape variety and vintage can be very compressed, with all the grapes coming in at the same time.

Pernod Ricard, owners of Brancott Estate, have adopted continuous flotation. They remove the clear juice, and the floats then go through centrifuges to get any remaining juice out. The advantage is that within 24 hours everything is inoculated and ready to ferment. This is the key to getting clean wines in vintages such as 2017, when there was quite a bit of botrytis around. Flotation is quick, so the juice only has limited time in contact with any of the botrytis products that might be in the must.

Villa Maria, another major player in the region, also use it. "Yes, we have a continuous float unit for Sauvignon Blanc juice," says the chief Marlborough winemaker Helen Morrison.

"These days we use just nitrogen gas and small dose of bentonite." Villa Maria also has a centrifuge. Natalie Christensen, chief winemaker at Yealands, says that they do both continuous and batch flotation. "For continuous float we use a bentonite called Flottobent at a super-low rate (0.36ng/l) along with nitrogen," she says. "And for batch floating we use only nitrogen."

JUICE STABULATION

Juice stabulation is a newly popular winemaking technique first used by winemakers in Gascony a couple of decades ago. However, it is its recent adoption by rosé winemakers in Provence that has given this method more attention. Their rosé wines have become progressively paler over the years, and this has presented winemakers with something of a challenge. Many of the flavor compounds in red wine come from the skins, and for rosé this is also true, albeit to a lesser extent due to the short time it is left in contact with the skins (just long enough to get the desired color). So, if the wines are getting paler, the length of time in contact with the skins is getting shorter, and the extraction of flavor compounds is getting lower. This results in wines with less flavor, so Provençal winemakers have come up with a technique to offset this, called juice stabulation.

There exists a research institute dedicated solely to the rosé wines of Provence, called the Centre de Recherche et d'Expérimentation sur le Vin Rosé, based in Vidauban. Among other things, they have devised a color scale for rosé. "The fashion now here is to have very pale rosés," says Véronique Goupy of Domaine de Fontlade, "But the aroma is in the skin. If you don't macerate enough you have no aroma and no color. It is a very technical wine." This means winemakers are walking a tightrope between getting aromatics in their wines and yet keeping the color pale.

One way to improve aroma without extended maceration is stabulation. Aurélien Pont of Château Pigoudet explains how this works. "For the last six years we have left the juice on the lees for five–fifteen days before fermentation at low temperatures. This allows lots of aromatic components to go from the lees to the juice." Pont adds that, "If you make a tank just from filtered lees it is very aromatic and can be used as a blending component."

The key to stabulation is extracting flavor—and flavor precursors—from the gross juice lees (not to be confused with the yeast lees that form during and after fermentation), without exposing the wine to oxygen, and without extracting any bitter flavors. Normally, after pressing the juice lees are removed from the wine by sedimentation (also called settling, which is usually for 24 hours, followed by removing the now-clear juice off the lees), or by flotation (see above). But here the intention is to keep the juice in contact with the lees long enough to extract any flavor precursors that might be present. It is important that fermentation does not start, because, once it does, clarifying the juice is not possible and you end up with a full-solids fermentation, with an attendant risk of reductive characters developing. So, during stabulation the juice is kept cold: some use temperatures as low as −2°C (28°F); while others might go as high as 10°C (50°F), although this is a bit risky. It is also stirred up, but without introducing oxygen, so the stirring will be done with carbon dioxide, or by adding dry ice, usually once a day. This also sparges out any dissolved oxygen.

The length of time for stabulation can vary, but it is usually four days to two weeks. Some oenological product companies offer enzymes that can be used to help release aroma precursors, and these can reduce the time needed for stabulation. Winemakers are aiming to get increased levels of thiol precursors from the lees, which the yeasts then turn into the aromatic polyfunctional thiols. These are important in the aroma of Sauvignon Blanc, too, and some Marlborough wineries are trialing stabulation as a way of increasing their concentration in the wine. They add a nice fruity component to rosé. The other compounds that stabulation is supposed to increase are the esters, which give fruity aromas, and which are sensitive to oxidation. Stabulation is also reported to improve the mouthfeel of the wine.

Are there any drawbacks? For rosé, if the contact is extended too long, then there can be some color loss. Also, because of the repeated suspension of the solids in the wine, it can be difficult afterward to get the required degree of clarity for fermentation by settling. Centrifugation is one option that avoids this problem, and flotation is another.

17 The science of sweet wines

When it comes to making sweet wines, there are a number of strategies that can be used. Grapes contain a lot of sugar: at harvest their sugar content will typically be around 240g/l. During fermentation, yeasts then convert this to alcohol and carbon dioxide. Around 17g (0.6oz) of sugar gives one degree of alcohol, although the alcohol yield varies a little with fermentation conditions and yeast strain. It follows that as fermentation progresses, the wine gets drier. To make a sweet wine, fermentation must be stopped early, or sugar levels in grapes must be very high, or the sugar levels in the grapes must be concentrated or added to in some way. Here I'll explore the science behind sweet winemaking, beginning with one of the most intriguing methods, noble rot.

BOTRYTIS: "NOBLE ROT"

I don't know who first came up with the idea of making wines from rotten grapes. It sounds like quite a bad one. Perhaps it was necessity? You could imagine someone thinking, oh no, my grapes are rotten, but let's try making something from them and see how it turns out. Or it could have been a discovery made by someone in desperation gleaning the rotten grapes that had been left on the vine. But whoever first took the plunge, I am glad they did, because sweet wines made from grapes that have succumbed to "noble rot" are some of the treasures of the wine world.

TOKAJ

It is October 2013 and I am in Hungary, in the vineyards of Tokaj, at harvest time. Ever since I heard about Tokaji, the almost mythical sweet wine from the Tokaj region in Hungary, I'd been fascinated by the concept. While Sauternes in Bordeaux is probably the most famous of the wines made from noble rot, it is just a small footnote in the regional story of Bordeaux.

But the sweet wines from Tokaj are the main focus of this region, and have a long and illustrious history, dating back 1,000 years. King Louis XV of France famously described the aszú wines of Tokaj as the wine of kings and the king of wines. And in the Hungarian national anthem there is a line about the nectar—the Essencia of Tokaj.

I am visiting László Mészáros at Disznókő, a wine estate in the region owned by French wine group AXA Millésimes. He has been working here since 1995, and in 2000 was put in charge. He gives a rundown on the geography of the

Above A bunch of Furmint grapes, some of which have been attacked by botrytis. These shriveled and sometimes raisined berries are soon to be picked, leaving the healthy berries on the vine behind for a separate picking stage.

region, the vineyards of which are situated on the eastern and southern slopes of the Zemplén hills. "The eastern border is a river called Bodrog, which joins the Tisa, a larger river crossing the east of Hungary," explains Mészáros. "They join in the town of Tokaj. In the south we have the Hungarian plane. The region covers around five and a half thousand hectares of vineyards, in a sort of V shape, covering an area around fifty by thirty kilometers [31x19 miles]."

There are 26 villages in the region. "After the first world war, two and a half villages in Tokaj became part of Czechoslovakia and this is why there is a small amount of Slovakian Tokaji," he says. The bedrock in the region is volcanic, but over this base there are diverse soil types. "There are two main types: the loess soils, which are mostly around the Tokaj hill, and the clay soils that form the majority of the wine region. Sometimes the clay is a very heavy, black clay. Often the clay is mixed with loess, or volcanic pebbles."

We go into the vineyard to watch the harvest, which takes place in stages. The first few passes are focused on picking just the shriveled, nobly rotten berries out of bunches that are otherwise intact. It is an incredibly slow, precise process. Mészáros shows us some of them: it's interesting that while there is evidence of botrytis on some, others appear to be simply dehydrated, like raisins. They taste incredibly sweet and rich. One person harvests around 6–10kg (13–22lb) of aszú berries a day.

He explains that the seeds should be brown. If they are not, he won't make an intense maceration with them. The seeds on these berries are quite green, and they can contribute too much tannin if they are macerated for a long time. The most important grapes here are Furmint and Hárslevelű, and they represent around sixty-five percent and twenty percent of the region's vineyard area respectively. Four other grapes are allowed to be used: Sárgamuscotály (aka Muscat Blanc à Petit Grains), Kabar, Kövérszőlő, and Zéta.

Above Aszú berries, selectively picked in a pass through the vineyard. These are kept separately in a special tank, and later form the aszú paste that is added to fermenting wine to give sweetness and complexity.

You can't use these aszú berries to make wine in a conventional way because there simply isn't enough liquid. So, the winemaking process is complicated and quite unique. The first step is to collect the aszú berries in a tank. These berries are picked throughout September and October and they can be stored until it is time to use them: with their high concentrations of sugar and acid, they aren't going to start fermenting in a hurry. Sometimes they are even stored as long as two months before use. These tanks release a small amount of highly concentrated juice, with sugar levels of 600–900g/l. The juice is drained off into demijohns. The lighter ones will be added to the sweet wines after fermentation, but the heavier ones will become Essencia, an incredibly rare, sweet wine. It barely ferments at all, and final alcohol levels might be 4%. But it is astonishing stuff, with the intense sweetness balanced by the acidity.

The key stage in the winemaking process is the maceration. The aszú berries are soaked in fermenting wine, or wine that has just finished fermenting. The period of this soaking process depends on the wine and the quality of the berries, but usually it's from 12–60 hours, and involves both pumping over the wine and punching the cap down, to extract as much of the sugar and complex flavor from these little sugar bombs as desired. Then the juice is drawn off, and the mush of aszú-berry skins is pressed slowly, with the pressings added back to the wine. Fermentation can then carry on until the right balance is reached. The Tokaji aszú wines are graded on a scale of puttonyos. In the past, this designated how many *puttony* (hods of 27l/7 gallons of aszú berries) were added to a typical barrel of wine from the region (these *gönci* casks were 136l/36 gallons in capacity). Now it refers to the residual-sugar level of the wine. The number of puttonyos used to range from 3 to 6, but now the lower tier has been removed and they are 4, 5, or 6 puttonyos.

BORDEAUX

With my appetite whetted by the visit to Tokaj, the following year, I am in Bordeaux, visiting Sauternes and Barsac, and my first appointment is with Prof. Denis Dubourdieu. He was one of the most important figures in developing the white wines of Bordeaux, both sweet and dry. I had caught up with him in 2014 to interview him: he was then 65, and sadly died in July 2016, far too young, at the age of 67. Denis was primarily a research scientist and spent half of his time working for Université de Bordeaux (where he was president of the Institut des Sciences de la Vigne et du Vin) and half consulting and working on his own project, Domaines Denis Dubourdieu, which consists of five estates. We met at Château Doisy-Daëne in Barsac, where he was born. His consultancy business was significant and involved 70 clients, one-third of whom were in Bordeaux, and as well as Denis, the team included Valérie Lavigne and Christophe Ollivier. "I have three lives," he told me, "but they are not so far away from each other: I teach what I do, I do what I teach." It is his contribution to what we know about the flavor of white wines, both dry and sweet, which will probably have the biggest legacy.

Famously, working together with Takatoshi Tominaga, Denis Dubourdieu discovered the role of a group of sulfur-containing compounds, the polyfunctional thiols, in the aroma of white wines. "It was the discovery of my life," he told me. "No one could believe that thiols could be involved in the aroma of grape varieties, and that S-cysteinylated compounds were the precursors." He discovered many things, including the influence of the yeast turning these precursors into thiols during fermentation.

These thiols, particularly 3MH, 3MHA, and 4MMP (see page 125) give the distinctive passion fruit, grapefruit, and boxwood aromas that are typical of Sauvignon Blanc and botrytized wines, but which also occur in other grape varieties. This research has been followed up elsewhere, particularly in New Zealand, where Sauvignon Blanc from Marlborough show particularly high levels of 3MH and 3MHA. How do you get good levels of these nice thiols in white wines? "The level of the precursor is important and the phenolic content." Press juice has high levels of both, and the wines are often disappointing because the precursors end up being oxidized. A good level of nitrogen in the juice is helpful, i.e. 200mg/l FAN (free amino nitrogen, which together with ammonia makes up what is known as YAN or yeast assimilable nitrogen—the nitrogen that is available to the yeasts). So, the ideal juice would have low pH, high precursors, and low phenolics.

Other researchers have identified further polyfunctional thiols that contribute to the aroma of nobly rotted wines. In 2006, Bailly and colleagues from Belgium identified two thiols that give these wines a bacon/petroleum aroma, and which work synergistically together. They also showed that a range of compounds contribute to the aroma of Sauternes, including α-terpineol, sotolon, fermentation alcohols (3-methylbutan-1-ol and 2-phenylethanol) and esters (ethyl butyrate, ethyl hexanoate, and ethyl isovalerate), carbonyls (trans-non-2-enal and β-damascenone),

Above Nobly rotted grapes in Sauternes, Bordeaux. The fungus, *Botrytis cinarea*, infects ripe grapes when the morning mists encourage its presence. It shrivels the berries, concentrating sugar and acidity, and the altered berry metabolism produces interesting flavor compound precursors.

and woody characters (guaiacol, vanillin, eugenol, β-methyl-γ-octalactone, and furaneol). In a paper published in 2009, the same group looked at the effects of aging, and showed that after two years in bottle, the only thiol left above its threshold value was 3MH. The other aroma compounds survived in the bottle for at least six years (the furthest they looked), and the paper also mentions another three aroma compounds that have been newly identified exhibiting interesting cake, honey, and dried-apricot characters: homofuraneol, theaspirane, and γ-decalactone.

When it comes to botrytis, it is important that the fungus acts on healthy, ripe, metabolizing grapes. They need to respond actively to the infection by the fungus, and there should not be too much time between ripeness and infection. "Infection is not only about concentration," Denis said. "It is stimulation of the production of aroma precursors by the pulp. When the grape becomes an old guy it is difficult to be excited by the fungus." More mature grapes simply concentrate their flavors, with honey and citrus, when they are infected. When grapes that are metabolically active

get infected, then you get the polyfunctional thiols, with flavors of apricot, mango, and grapefruit. "You want fast ripening and fast invasion."

In September 2016, I visited again, this time to spend time with his son Jean-Jacques Dubourdieu, who now runs Domaines Denis Dubourdieu. For Sauternes, the "better" vintages in Bordeaux are not always the best. "In the bigger vintages with more concentration you get more volatile acidity, approaching one gram per liter, and the wines don't always age as well," says Jean-Jacques. Winemaking in Sauternes is a little simpler than that in Tokaj. The grapes are often picked in one pass through the vineyard (although some estates will do multiple harvests) and in a good year on a good site there will be lots of botrytis. But often there are quite a lot of clusters that are only partially botrytized, or even not at all. The botrytis doesn't just add the exotic flavors and the sugar; acid levels are also higher because of the concentration that comes through dehydration. The key to great sweet wine is the balance between the sweetness and the acidity—both should be high. Then the grapes are pressed, releasing clouds of spores as the press is loaded, and the juice is transferred to barrel for fermentation. Some new barrels may be used, as the vanilla flavor from new oak meshes quite well with the rich flavors of Sauternes and Barsac.

GERMANY

Germany has a great tradition of making wines with botrytis influence, and the hot spot for sweet wine production here is the Mosel. This is one of the world's most spectacular wine regions, with its steeply sloped vineyards running along the Mosel River, and two of its tributaries, the Saar and Ruwer. Riesling is king here and occupies over 60 percent of the vineyard. From this, a wide range of wines are made in different sweetness levels from bone dry to lusciously sweet. Different vineyards have different talents. Some might primarily be dry-wine specialists while others are better at sweeter styles. But, typically, the vineyards will be selectively harvested and wines of two sweetness levels in particular, BA (beerenauslese) and TBA (trockenbeerenauslese), are made exclusively from nobly rotted grapes. These are some of the most complex, intense, and expensive wines of all. One of the great advantages of the combination of the Mosel and Riesling is the naturally high acid levels. For compelling sweet wine it is important to have a balance between sugar and acidity and in the Mosel this works perfectly. Germany is also home to another method of making sweet wine by concentration: ice wine.

ICE WINE

Ice wine is incredibly sweet, but it also has amazing acidity. It is made by picking frozen grapes, which then yield a super-concentrated must as they begin to thaw in the press, with massively elevated sugar and acidity.

And while ice wine originated in Germany and Austria, it is now Canada that leads the way. Head to Toronto and drive south for an hour or so on the Queen Elizabeth Way, and you come to the Niagara wine region. That this sizeable vineyard area exists at all is because of the lake effect: Lake Ontario keeps things cool during the frequently hot and humid summers, and because it is deep enough that it rarely freezes, it stops winter temperatures dropping too much in the winter. *Vitis vinifera* cannot survive in temperatures much below -18°C (-0.4°F), and in the rare winters where the lake does freeze, such as 2013/14, then some vines are going to die, especially the less cold-hardy varieties. Ice wine represents just 0.3 percent of export volume of Canadian wine, but because of the high prices it fetches it represents one-quarter of export value. Ontario rules the roost here, although some is also made in British Columbia. Ontario alone produces 80 percent of the world's ice wine.

Ice wine—as in wine from frozen grapes—was first made in Roman times, although details of the winemaking are not clear. After this, the first record of ice wine (or Eiswein) comes from Franconia, Germany, at the end of the 18th century, and again, details of this are sketchy. The first detailed records are from 1830, also in Germany, where sweet wines from botrytis-affected grapes were already being produced. In subsequent decades, this remained a rare style of wine, made only when winters were particularly harsh and early. Things picked in the 1960s when the availability of temperature alarms (alerting winegrowers to sufficiently low temperatures) and electric lights to aid picking during the early hours of the morning, helped make Eiswein a much more regularly produced style. But over the last decade and a bit, warming trends have made good ice-wine vintages a rarer occurrence, and production has therefore been less widespread. In Germany and Austria, Eiswein remains an exotic treat.

Step forward Canada. The first Canadian ice wine was made in the Okanagan Valley, British Columbia, by ex-pat German Walter Hanile in 1972. This was just a tiny amount—40l/10.5 gallons—and he ended up not selling it until six years later in 1978. But the Canadian ice-wine story really began in Ontario. In 1983, a group of four wineries with German/Austrian influence left grapes on the vine with a view to pressing them frozen: Karl Kaiser of Inniskillin, Ewald Reif, Hillebrand, and Pelee Island. The entire production of Inniskillin and Reif was lost to birds, but the other two producers harvested small quantities. The next year Kaiser used nets and made an Inniskillin ice wine in 1984 from Vidal grapes.

Most of the momentum for the category came from Inniskillin. "Our first ice wine was made in Ontario in 1984," says Derek Kontkanen, Winemaker with Inniskillin Okanagan. "Karl Kaiser, our founder, was from Austria and he knew how to do it. Back in 1991, we won the Grand Prix d'Honneur at 1991 Vinexpo [in France, with 1989 Vidal ice wine] and this kind of put Canada on the map." Now, Inniskillin ice wine is sold in 74 countries around the world.

For the larger companies in the Ontario wine scene, such as Inniskillin, Jackson-Triggs, Peller, and Mission Hill, ice wine is a significant earner. Marco Piccoli is winemaker at Jackson-Triggs, one of the main players in ice-wine production. I asked him whether ice wine is important for his company. "Absolutely," he replied, "for different reasons. First of all, it is what put Canadian wine on the map. Second, I can claim that Canada is one of the best at making this kind of wine. We can really be competitive with many other sweet wines. Third, it is a very profitable wine."

One of the keys to Canada's success with ice wine is that it is able to make it consistently, and Piccoli thinks this is important. "I can make very good Cabernet Sauvignon, but can I do that every year? No, of course not. But I can do very good ice wine every year." The quantity can vary though, depending on the timing of the onset of the cold part of winter. "Some years we make two hundred thousand liters [53,000 gallons], in other years just twenty thousand [5,300 gallons], but the average is probably forty to fifty thousand liters [10,500–13,200 gallons]."

There are viticultural challenges to making ice wine, and the first is managing the fruit load. "You don't wake up in the morning during harvest and say, those grapes are for ice wine," says Piccoli. "You need to prune them differently and keep the vines loaded a bit more. You want the grapes to ripen later so as they go into the cold season they don't rot. They need to be protected from the birds and animals. Harvest is not that big a challenge because as long as it is cold you pick." One of the advantages Canada enjoys is that the extremely cold winters mean that temperatures will always drop low enough to facilitate its production. Legally, to make ice wine in Canada you need temperatures below -8°C (18°F), whereas in Germany and Austria the rule is they must be below -7°C (19°F). The optimum is between -10 and -12°C (14 and 10°F), because then the grapes are cold enough that

minimum sugar levels are achieved, but not too cold that hardly any juice is yielded. For German Eiswein the minimum sugar level is 110 Oechsle; in Austria it is 125 Oechsle; and in Canada it is 35 degrees Brix (153.5 Oechsle). The freezing process concentrates the sugars in the juice and also concentrates acidity, which is essential to balance the sweetness in the final wine. The grapes must be picked and pressed in the same night, without ever warming up beyond the minimum temperatures specified. The grapes are normally pressed in a basket press, and many wineries have pressing stations out in the vineyard so that they can be processed quickly without warming up.

Vitis vinifera grape varieties, such as Riesling and Cabernet Franc, are picked the first time they get cold enough. But Piccoli treats Vidal, a hybrid, differently. "It is typically very phenolic, so we want to leave it to freeze and thaw for a couple of cycles. This allows the phenolics in the skins to oxidize naturally. When you press the grapes the phenolics are already oxidized and you have less work to do."

"From the winemaking side, yeasts do not like the sugar," says Kontkanen. "It is a very high sugar content for them and they struggle. We end up with higher volatile acidity than you find in table wine." He targets 42 Brix in the juice, with acids in the range of 10–15g/l. Fermentation takes from ten days to two months. Kontkanen likens the experience of yeast cells in the must of ice wine to a human diving at such a depth that the pressure is crushing. "If you look at a picture of the yeast cells in the must, the yeast cells are actually crushed, so they produce a lot of glycerol internally to regulate this, and they spit out volatile acidity as a by-product."

STRAW WINE AND DRYING GRAPES

STRAW WINE
What happens if you grow grapes in a warm climate and want to make a sweet wine? One option favored by Mullineux wines in South

Africa's Swartland wine region is to make what is called a straw wine. "When we moved to the Swartland in 2007, we thought about what kind of wine we wanted to make here, and when it came to making a sweet wine, there are a few different options," explains Chris Mullineux. Ice wine was not an option in the Swartland, and they don't have the humidity for botrytis to make noble late harvest. Late harvest is an option, but it is hard to have the acid needed to balance the sweetness when you make a wine this way (see page 201). "For us, straw wine was the obvious way to make a sweet wine in our climate."

It all begins with viticulture. "In the Swartland we have warm, dry, breezy conditions so there is very little disease pressure," says Andrea Mullineux. "We have tried to maintain the healthiest fruit at harvest. The ideal for us is that in the winemaking process when we are air drying the grapes we are concentrating the sugars and the flavors and, most importantly, the acidity."

Their straw wine is sourced from the same dry-grown old-vine Chenin they use for their dry Chenin Blanc. They capture freshness by harvesting at normal ripeness. "By harvesting them at normal ripeness, we are stopping the ripening process by cutting the grapes off the vine," says Andrea. "We are not losing the acidity in the drying process. One of the most important things is the shape of the bunch and the size of the berries. We don't want bunches that are too compact, or too loose."

With late-harvest wines, the ripening process continues and acidity is lost as sweetness rises. It is only really an option when you start out with very high levels of acidity, such as Riesling in the Mosel. With the straw-wine concept, the acidity is maintained and even rises as the sweetness increases through drying. They dry the grapes outdoors under shade. The bunches are laid out one layer thick, and the process takes two to three weeks. Over this time the moisture evaporates and the grapes start to shrivel. This concentrates sugar, flavor, and acidity. "In the vineyards we are

normally getting four to six tons per hectare," says Andrea. "We usually get seventy-percent recovery from that, so from every ton of grapes we get about seven thousand liters [nearly 2,000 gallons] of juice. With the straw wine we get a ten-percent yield. So, for every ton of grapes we only get about a hundred liters [26 gallons] of juice."

Then it's time to press. "It takes two days of long, slow pressing," says Andrea. They select from the juice: some is just too sweet, and sometimes not sweet enough. Press fractions are taken. The juice is so viscous it can't be settled. But because it is a long slow pressing, the juice is coming out fairly clear. It goes straight to barrel and there is a natural fermentation. This kicks in pretty quickly, even though the juice is very sweet. "A really strong yeast culture has built up on the grapes," says Andrea, "so they start fermenting within a couple of days." Then the fermentation takes up to ten months: the fastest is seven months. "We don't stop the fermentation," she says. "For us it is important that it naturally stabilizes. To stop fermentation you would have to add a lot of sulfur dioxide, or freeze it, or sterile filter it." There are variations in the sugar, acid, and alcohol each year, but it maintains the same ratio. In sweeter years the acid is higher, which is ideal.

For Mullineux' vintage straw wine one new barrel is introduced every year, but this means that the majority of them are older barrels. The wine is left until it is naturally stable and is then bottled as a vintage straw wine every year, usually at vintage so the barrels can be filled with the next wine right away. These are not barrels you want to leave empty. Then each year they keep two barrels back to make a solera for fractional blending across vintages to make their Olerasay wine. In the solera, the wine stabilizes. As the wine ages, it develops a dark amber color. But anything that can drop out, including the amber color, will drop out in this aging process. When the new vintage of wine is added to the solera, it brings it back to life. "This solera, which spans the vintages 2008 to 2019, tastes younger than any of the individual vintages did on release," says Andrea. "This is the beauty of the solera system."

Mullineux was not the first in South Africa to make a straw wine: that credit goes to De Trafford, which first made its excellent straw wine in 1997, inspired by the *vins de paille* of the Northern Rhône in France. In addition to its straw wine and Olerasay, Mullineux make a third sweet wine, called Essence, made from the very last "hard" pressing of the straw-wine grapes, which is incredibly rich in sugar and acidity. The 2012, the first release of this wine, had so much sugar that fermentation took five years and then stopped at 4% alcohol. The remaining sugar level is 680g/l, the acidity an astonishing 15g/l. It resembles the Essencia wine made from the free-run juice from the aszú berries in Tokaj.

DRYING GRAPES

The Veneto in Italy also has a tradition of drying grapes, but this is largely for making more-or-less dry Amarone wines. This drying process is known as *appassimento*. One of the key producers in the region, Masi, has invested in research on the *appassimento* technique, and have a research group headed up by their chief winemaker, Andrea Dal Cin, who has worked with Masi since 2002. Dal Cin explained more about the technical details of *appassimento*. "Something special happens inside the berries," he says. "Genes that are normally switched off suddenly switch on after sixty or seventy days of drying." They make a careful selection of the grapes and place them on bamboo racks, not plastic. The bamboo works better: if a grape leaks, the juice is soaked up. It is possible to increase the speed of drying, but then the grapes taste different. "We prefer ninety-to one-hundred-and-thirty day *appassimento*."

Dal Cin also likes a bit of botrytis attack of around five–ten percent, but not when they receive the grapes: they want it to develop on the grapes during the drying process. "Botrytis produces

two enzymes that can oxidize the wine," he says. "Tyrosinase and laccase are oxidative, attacking the color and oxidizing the aromas." They want the botrytis to occur on the Corvina grapes only, at a rate of five–ten percent. The result of this drying process is: 30–40 percent loss of weight; the concentration of color, sugar, tannins, and acids; some glycerol from the botrytized grapes; and evolution from primary to complex aromas.

Masi has its own strains of yeast—called Masy-03, Masy-04, and Masy-0—that have been selected from dried grape fermentations. These ferment at a very low temperature, beginning at 5–8°C (41–46°F) and then fermentation lasts for 35–40 days. "Fermenting really cold improves the glycerine, which adds to the glycerine produced by the botrytis attack," says Dal Cin. Glycerine adds to the mouthfeel of the wine, making it seem more richly textured.

LATE HARVEST

South Africa is also home to one of the world's great historic sweet wines, which came to fame in the 17th century, and which for a while was discontinued before being reborn. It is the Vin de Constance, from the Constantia wine region. In the 18th and 19th centuries, Constantia was one of the world's great sweet wines, but it disappeared with the phylloxera crisis in the late 19th century. Since 1986, though, the Klein Constantia winery has been making a sweet Constantia that it labeled Vin de Constance. It is based on the original wines that were made here. When the Jooste family bought the property in 1980, they were approached by a Stellenbosch Univeristy viticulturist, the late Prof. Chris Orffer, who encouraged them to try to recreate the historical sweet wine. They agreed, Orffer helped them, and through the skill of the winemaker at the time, Ross Gower, they succeeded.

The historical Constantia wine is made from allowing the grapes to stay on the vine for a long time and achieve very high levels of sugar, in part through desiccation and raisining of some of the berries. It was likely made from a grape variety called Muscat. So, for the sake of historical accuracy, Vin de Constance is a late-harvest style made from Muscat without botrytis. This variety had disappeared from the vineyards, so it was replanted in 1982, and the first new-era Constantia was the 1986 vintage, released in 1990.

Normally, late-harvest wines can be a little simple and can lack acidity, which is lost as the grapes hang on the vine accumulating sugar. So, the winemaking team had to work out a special protocol for making a complex, world-class late-harvest wine. First, ten percent of the grapes are picked early, to make a base wine at 12.5–13% alcohol with good acidity. Then they go and pick any already-raisined grapes. The remaining grapes are left to accumulate sugar, and leaf removal exposes the fruit to the sun, which helps more of them to raisin. Then, the big pick takes place in three passages through the vineyard. Altogether, around ten percent of the crop will be raisins. The key stage is extracting the flavor and sugar from these shriveled berries, and to do this some of the first-made base wine is added to the raisined berries, which then slowly release their flavor. The skins are pressed quite hard, because tannins from the skins, normally a key part of red wine flavor rather than white, are important to the style, adding freshness to counter the sweetness. The wine then spends about four years in barrel until it is clear and ready to bottle. The result is a very complex, very sweet wine with astonishing aging potential.

The original Constantia estate was huge, and a long time ago was split into a number of different properties. Since Klein Constantia released its Vin de Constance, neighboring estates have joined the game: Groot Constantia followed in 2003 with its Grand Constance, and Buitenverwachting has its 1769. On my last visit to Klein Constantia, as well as tasting the brilliant 2013 Vin de Constance, I got to try a bottle of the 1875 vintage of this wine, which was one of the last years of the sweet wine being made here before its production stopped. This bottle had been brought in by a collector, and the team at the winery reckoned this wine was probably bottled in Europe from a barrel that had been shipped there. They took the chance to analyze this piece of vinous history. It weighs in at 14.7% alcohol and had 220g/l of sugar. We sipped the small pour we were given in a state of humble appreciation. It had aromatics of leather, spices, old furniture, as well as raisins and table grapes with some honey, too. It was honeyed and smooth on the palate with real harmony, and notes of barley sugar, grapes, stewed raisins, bread pudding. Remarkable.

FORTIFICATION

One way of making sweet wines is to pick grapes at normal-ish levels of ripeness while they still have good acidity, and then stop fermentation while there are still high levels of sugar by adding high-proof grape-derived spirits—young, colorless brandies. This is called fortification. And the interesting thing about fortified wines is that while most sweet wines are white, the most famous fortified wine of all is largely red: Port.

Port is a fortified wine from the spectacular Douro wine region in Portugal, and it owes its existence to trading relationships with the English in the 17th century. Vines have been grown in the Douro since as far back as Roman times, but this was a poor region and grapes were grown as part of a polyculture: the horizontal areas of the terraces were reserved for food crops, while vines were planted in the gaps of the terrace walls. The English and Portuguese developed a special trading relationship that flourished off and on for a number of centuries. The Douro's big break came when war broke out between France and England in 1689, which forced the English to tap new, non-French sources of wine. This led to an increased demand for the newly discovered Port wine. In those days doctoring wine was commonplace, and it was found that the addition of brandy had a twofold benefit: it made the wine more stable, helping it survive the voyage to England without harm, and because the brandy was added before fermentation was completed, it made the wine sweeter. The English developed quite a taste for it.

One of the advantages of fortification was that it made the wine much more stable, early on in its life. So even a small quinta in the Douro could make a Port of good quality without access to technology or winemaking expertise. They then sell it in barrel to the Port houses for transport down river to Vila Nova de Gaia where the conditions are much better for wine storage and aging than in the hot summers of the Douro.

To make Port in the traditional way, grapes are harvested and then put into shallow stone fermenters called *lagares*. In the past, these would have been foot-trodden, and some still are, quite intensively for two–three days before the spirit is added (see page 146). The advantage of the foot is that this allows for intense extraction from the skins without breaking the seeds. Many Port houses now use robotic treaders to simulate the human foot, because workers to do this task are in short supply. Fermentation kicks in but is then stopped by the spirit after about half the sugar has been used up: the typical residual-sugar level in Port is around 100–120g/l. The fortification takes the alcohol level up to around 20%, which removes the risk of microbial growth and problems like *Brettanomyces* and volatile acidity. Because wine relies on fermentation for its complexity,

young Port wines can seem a little less complex and more fruity than table wines, which is why aging is so important for the top wines.

There are two styles of Port: ruby (where the wine is bottled young and aged in bottle, as with Vintage Port, Single Quinta wines, and Late Bottled Vintage) and tawny (where the wine is aged in casks and barrels for many years; vintage-dated tawnies are known as Colheitas).

Fortification is used elsewhere to make wine, most notably in the Sherry triangle in Spain's Andalucia (most of these wines are dry, fortified after fermentation has finished) and in the production of Vins Doux Naturels in the south of France. For many of these, fortification occurs before very little fermentation has taken place. Where fortification occurs after only a limited fermentation, the wines lack some of the complexity that comes from yeast metabolism, and instead the focus is on the pure, fruity flavors from the grapes.

THE FUTURE

Sweet wines have historically been among the most sought after of all fine wines. They can be thrillingly complex. But are these great wines simply historical artifacts, or do they have a future? This discussion is beyond the scope of a book on wine science, but the interest these wines attract suggests that while selling huge volumes of sweet wine—especially fortified wines—will remain a struggle, the market has a place for smaller volumes of high-quality sweet wines.

18 Taking account of individual differences

It is a Friday morning in Adelaide, and I am at the Australian Wine Research Institute (AWRI), tucked into the corner of a campus on the outskirts of the city. I am here to meet with senior sensory scientist Wes Pearson, who is busy at work in the preparation room of the institute's sensory laboratory. This is a busy place: the AWRI run frequent sensory sessions, both with consumer and expert panels. Here they try to understand the way in which wine composition translates into wine flavor, and why some people prefer some wines to others. Pearson is juggling with a large array of vials, each of which contain individual chemical constituents of wines. He uses a special pipette to accurately spike a commercial red and white wine with a particular concentration of these chemicals, so that we can assess their sensory impact. This time we are looking at a range of fault compounds, among others.

Along one side of the preparation lab are small wooden windows. These each open into small sensory booths, which is where the panel members sit. The booths are isolated, and are a spotless white, with a small sink recessed into each desk. The idea is that the panel members should have no distractions at all and be free to concentrate on the assessment task they are faced with. The window opens from the next room, the glasses are passed in, and the panelists taste, marking their verdicts—usually in the form of a point on the line of an analogue scale for a number of previously agreed attributes—on a small tablet computer in front of them.

One of the samples Pearson prepares is of a compound called β-ionone, which contributes to the fruity aromas of some red wines. "I'm interested in whether or not you can smell this," he says, pointing out that a proportion of people just don't get it at all. I take a sniff of the doctored red wine, and immediately there's a big blast of floral/raspberry aroma. I can smell β-ionone. It turns out that there are quite a few compounds like this in wine that we differ in our ability to smell.

One of the most famous is rotundone. A few years back, the AWRI identified this as the compound responsible for the peppery character in red wines, especially cool-climate Syrah. As well as making some red wines smell peppery, it is also responsible for the peppery aromas of peppercorns. Surprisingly, though, one in four of the population cannot smell it at all, even at heroic concentrations. It turns out that ionone and rotundone are just two of the growing band of wine-relevant aromas that we differ in our ability to detect.

ARE WE LIVING IN DIFFERENT TASTE WORLDS?

The concept that we might all be living in rather different taste worlds dates back some 85 years. It was in 1931 that A. L. Fox, working at the DuPont industrial research laboratory in Wilmington, Delaware, made a momentous discovery. He had been synthesizing a novel chemical and knocked over a vial full of it. As the dust flew into the air, his colleague remarked on how intensely bitter it tasted, while Fox couldn't taste it at all. Health and safety rules were different then, and Fox ran around the building testing whether people could detect it; 60 percent could, while the remainder found it tasteless. The chemical was phenylthiocarbamate (PTC). Propylthiouracil (PROP), which is closely related and somewhat safer, is now used instead and separates the population into groups of tasters and nontasters.

In the 1990s, researcher Linda Bartoshuk coined the term "supertaster" to describe a subset of tasters who have heightened sensitivity to PROP/PTC. This became an active field of investigation, as researchers showed "supertasters" seemed to have greater sensitivity to a broad range of different flavors, such as the burn of alcohol and the bitterness of coffee. Bartoshuk describes them as living in a "neon" taste world. While it was found that the different sensitivity to PROP was genetic and depends on the variant of the TAS2R38 gene (which codes for a taste receptor) that people possess, there also seemed to be a correlation between supertasting and having more tastebuds on the tongue.

Could it be that supertasters make better wine tasters? Initially there was a lot of excitement in the wine world about PROP-taster status and whether different sets of consumers might prefer different types of wines. And some leading critics came out claiming to be supertasters.

In 2012, John Hayes of Penn State University and Gary Pickering of Brock University, in Ontario, Canada, did a study that looked at a group of wine consumers and compared them with a group of wine experts. They found that wine experts are more likely to be medium- or supertasters than wine consumers. "If you have an innately superior ability to taste things, maybe you gravitate toward a field where that is rewarded," speculates Hayes.

But since then, new research has emerged that has made the PROP account look a little simplistic. Humans have 25 bitter-taste receptors, and TAS2R38, the one that detects PROP, is just one of them. There is TAS2R31, which detects the bitterness of artificial sweeteners acesulfame-K (AceK) and saccharin. It is possible to be highly sensitive to PROP bitterness and insensitive to saccharin bitterness, or vice versa. TAS2R31 is responsible for detecting the bitterness of quinine and also corresponds to liking for grapefruit. And TAS2R3/4/5 can explain the bitterness that some people experience with alcohol. "Bitterness is not a monolithic trait," says Hayes. "You can't measure one thing and expect it to predict everything."

Sticking with taste, there is a massive difference in the way people experience astringency. The mainstream explanation for the perception of astringency is called the delubrication hypothesis. Saliva contains proteins that lubricate the mouth, and tannins bind to these proteins and cause them to precipitate out. This strips the lubrication and we sense this change as astringency. But people differ widely in their salivary flow rate. The more saliva, the less astringent we find tannins. Also, people differ in their ability to replenish salivary proteins. As a result, some people are much more sensitive to astringency and like astringent foods and drinks much less. Hayes says that one-quarter of the population really to struggle to replenish their salivary proteins. For this group, many red wines will be unpleasantly tannic.

Then there are specific anosmias that are relevant to wine: odors that some people simply cannot smell. For example, the OR10G4 gene explains differences in responses to guaiacol, a smoky aroma that is found in wines suffering from smoke taint, but which is also present in some barrel-fermented and aged wines. The version of this gene that you have will predict how intensely you experience guaiacol, and those who experience it more intensely also like it less. Hayes thinks that this gene might also explain the different reactions of people to 4-ethylphenol and 4-ethylguaiacol, the two main aroma impacts of *Brettanomyces*, but that this work remains to be done. Variants in the OR2J3 gene influence how intensely people experience cis-3-hexanol, a grassy aroma. Variants in OR5A1 affect the way that the floral-smelling β-ionone is experienced (with a genetic association that means that roughly one-quarter of the population are insensitive to the violet/floral aromas found in some wines and will probably like these wines less). Plus, there is the common smell blindness to rotundone, mentioned earlier (see page 204), for which the gene is not yet known.

Another source of variation among people is in the enzymes present in the mouth, called glycosidases. Many aroma compounds in wine are found as inactive glyscosides. This is when an aroma compound has a sugar molecule attached to it and is no longer aromatic because it is now too heavy to be volatile. These glycosides are described as "locked away" aroma, and they are more abundant in wine than the free volatiles themselves.

Mango Parker, a researcher at the AWRI, has shown that people differ in the glycosidase enzymes found in their saliva. In a study looking specifically at this, she found that 6 out of 11 panelists seemed to have enough of these enzymes to liberate significant aroma from wine once it was in their mouths. So, it seems that variation in flavor perception is a little more complex that PROP-taster status alone.

"Yes, the PROP story has probably been oversold to the public," says Hayes. "It's an easy dramatic example that we really do live in different sensory worlds, but, no, it can't explain everything." He adds, "the molecular genetics of PROP tasting are actually very clear, but critically, they appear to be totally unrelated to 'supertasting.' In my view it seems very likely that PROP phenotype is capturing two totally separate sources of variance: receptor genetics (which we understand well) and overall amplitude/intensity (which we don't)."

SHOULD THE MARKET RESPOND?

So, should wine companies be taking this developing story of variation in flavor perception into account? "Yes, I think winemakers and producers should take individual differences into account," says Hayes.

"Probably not for PROP per se, but definitely for astringency or rotundone peppery aroma. I don't know of any that are doing it in a formal systematic way from the biology side." Hayes suggests that even smaller wine producers would benefit from acknowledging that not all consumers are the same. "It is critical because philosophically that would imply there is not one, and cannot be one, platonic ideal for 'the best wine,'" he says. "That is, those people who like sweet-fruity wines may not be uneducated philistines to be sneered at by wine enthusiasts; rather they may have a fundamentally different palate. And as capitalists, wine producers should be willing to meet the needs of these consumers, not just their winemaker friends and peers."

The degree to which wine companies could take this variation into account depends on how well it is characterized, and this is still early days in terms of doing this. Large wine companies have done extensive market research dividing consumers into target segments (great examples would be the Wine Nation and Project Genome segmentation exercises carried out by Accolade Wines [formerly known as Constellation Brands]).

However, these latest findings offer a way to segment consumers along biological differences, which could prove to be interesting. Rather than offer one wine designed to appeal to an "average" palate, it might be more effective to offer three that are specifically tuned to certain biological segments in the population. As such, a wine that has been developed to appeal to the average biology might end up pleasing no one.

"I think discrete segments probably exist, as they do for other food products," says Hayes. "But what I have spent the past eight years working on is trying to convince people that these segments are not merely idiosyncratic. Instead, they probably have a fundamental basis in biology. And if I am right, this would mean we can move beyond segmenting people for single foods, and instead generalize across foods."

IS TASTE PROGRAMMED?

There is another layer of complexity here. Our food-and-drink preferences are extremely malleable, for good evolutionary reasons. Innate preferences for nutritious, high-calorie foods are pretty universal. But we possess the ability to acquire novel tastes. Our sense of flavor combines with our memory, allowing us to explore unaccustomed, possibly nutritious food items, and these then become new flavor preferences (the memory aspect is important, because we need to reject quickly things that have made us sick in the past).

I like strong cheese, but 15 years ago I wouldn't eat cheese at all. The cheeses are still as pungent and extreme as they were 15 years ago. However, it is my response to the cheese that has changed, and I now love strong cheese. When I started drinking coffee as a teenager, I had it with two sugars and milk. Now if you put sugar in my coffee, I would find it unpalatable. And with wine, I began with richer, sweetly fruited reds that were easier to understand, and moved to more interesting wines over time. Today, I wouldn't find the wines I loved at first

all that interesting, and I wouldn't want to drink them. This sort of journey is not unusual. Who liked wine the first time they tried it? Most of the things I love now, when it comes to flavor, were tastes that I found challenging when I first experienced them. Experience has largely trumped biology, although biology can't always be overcome. "No amount of training can make up for fundamental biological differences," says Hayes.

There is one further factor to consider when it comes to matching people with their personalized wine: that of variety. The notion of testing my biology and then presenting me with the perfect wine for me is as problematic as suggesting you can match me with my dream meal. Much of my wine drinking is mood driven, and much of it is situational. I drink a wide variety of wines covering a whole spectrum of flavors, just as I eat wide range of foods. While I love the idea of data helping people find the wines they like, the whole idea is just far too simplistic, and does not take into account the complex factors involved in our choice of what we eat and drink. You aren't programed to like certain wines.

19 The future

Where is the wine world going? I'm no futurist, but to conclude this book I want to speculate about where wine science may be heading.

IN THE VINEYARD

There are changes afoot in the world's vineyards, with two looming issues that face viticulture as it is currently practiced. The first is climate chaos. Grapevines are sensitive to climate, but vineyards take years to establish, and choices of grape variety are often locally prescribed according to rules made when the climate was quite different to how it is now. Change is needed in response to the shifts in climate, but the viticultural world has a lot of inertia. Climate chaos will also increase susceptibility to certain pests and create issues around water use and availability. The second is sustainability, and here, I am referring to true sustainability. If a vineyard cannot be farmed the way it is now for the next one hundred years with no degradation of the soils, then that farming is not sustainable.

Most wine grapes are cultivars of a single species, *Vitis vinifera*. In the 19th century, it became exposed to two diseases and an insect pest that it did not coevolve with, and for which it lacks specific immunity. The consequence is that grapevines are exposed to huge doses of agrochemicals—both traditional and modern—and have to be grafted onto rootstocks. The environmental impact of all this spraying has been tolerated for too long. Instead, we need to look to the genetic resources of all grapevines and find new varieties that produce high-quality wine but without the environmental costs.

True sustainability without new varieties will be very hard, if not impossible, to accomplish in many locations around the world. Current breeding programs are offering some promising solutions, and many suitable varieties are already available, but the wine world remains wedded to just a small set of vinifera varieties.

Most viticulture has relied on a poor understanding of plant biology. One of the new areas of research has been looking at the unseen world under our feet: the soil. It is clear that there are very important interactions taking place in the rhizosphere that are critical to plant life and function. In the past these were just ignored. Weed control is a huge issue in vineyard management, and herbicides provide a cheap and tidy solution to this problem. But eliminating all plants other than the grapevine has massive implications for soil life. We need to consider the vineyard as an agroecosystem. Many problems currently solved chemically are no longer problems if the agroecosystem is allowed to find its own biological solutions.

Organics and biodynamics have only taken us so far, and we need a new, scientifically rational viticulture that is truly sustainable, and which can be applied intelligently taking account of local conditions. Regenerative farming and permaculture will feature increasingly. Plant signaling and communication will prove a fruitful area for study. Since Suzanne Simard's groundbreaking work in the forests of British Columbia, showing that trees share resources via underground fungal networks, the field of plant-microbe and plant-communication has become very interesting. It may turn out that vines are communicating with each other—and also the other nearby plants such as cover crops—through the soil microbial community. The contribution of vineyard microbes to the growth and success of vines will be unraveled, and there will likely be some surprises.

With drones, the cost of aerial imagery has gone down, and the use of multispectral imagery will make precision viticulture approaches more affordable. This will assist growers in their understanding of the heterogeneity in their vineyards, which can then lead to smarter management and better-quality grapes. Robotics will also become important in vineyards where labor is scarce or too expensive. Trunk disease is a major challenge, and part of the answer seems to lie in more intelligent pruning, and better nursery work.

IN THE WINERY

Winemaking is not chemistry. It is biology—and, specifically, microbiology. We are already seeing a lot of changes here, with growing use of non-*Saccharomyces* yeasts, and a greater understanding of bacteria and how they interact with yeasts. Next-generation sequencing has unlocked this hidden microbial world, and we will soon have a much better understanding of the microbiology in the vineyard and how this relates to what goes on in the winery. Bioprotection—the intelligent use of a range of microbes in fermentation—is going to become mainstream. It promises to cut the use of sulfur dioxide in wine. We are

going to see the emergence of a commercially significant category of "new natural" wines.

Intelligent winery design is coming, with super-efficient wineries that use less power and capture the carbon dioxide from fermentation. Green winery design is an important part of reducing the carbon footprint of wine.

With the increasing interest in artisanal, handcrafted, and natural wines, there will be growing use of concrete and clay in cellars, as well as large oak. Barrels will always have their role to play, but a smaller role than in the past. It would be great to have more of a scientific understanding of these alternative means of *élevage*.

In terms of perception, the old understanding of taste and smell—treating our olfactory and gustatory systems as measuring devices—is now outdated. The construction of the multimodal perception of flavor by the brain through interactions between different odorants and tastants is a complicated process. Gradually, the way this works is being understood. Having a good grasp of the nature of the perception of wine is important for a proper appreciation of what it means to "taste" a wine.

The world of wine is fascinating. It is incredibly diverse. There are many segments to it. People make, sell, and drink wines for different reasons. But, at its heart, this is more than a liquid in a glass with some pleasant flavor and mind-altering properties. It is culturally rich, intellectually engrossing, and provides the drinker with a connection to a time and a place. Science, used well, can help us understand and enjoy it better.

Glossary of wine science words

Abscisic acid An important plant hormone (also known as plant growth regulator), involved particularly in signaling during episodes of stress, such as cold or drought.

Acetaldehyde The most common **aldehyde** in wine, formed when oxygen reacts with **ethanol**. Present in small amounts in all wines. Not nice: smells bad, and one of the reasons that oxidized wines are not very pleasant. It's also the initial breakdown product of alcohol in the body, and responsible in large part for the unpleasantness of hangovers. Acetaldehyde is important because it is involved in the copolymerization of phenolic compounds. Combines readily with **sulfur dioxide**.

Acetic acid A volatile acid that is the main signature compound of vinegar. Formed by the action of *Acetobacter* bacteria on alcohol in the presence of oxygen. All wines have a little because it is a natural product of fermentation. But you don't want too much.

Acetobacter Bacteria that spoil wine by turning it into vinegar in the presence of oxygen.

Acids Important flavor constituents of wine. Provide tart, sour flavors that balance the other components. Wine contains a range of acids, most notably tartaric and malic acid (present in grapes) and lactic acid and succinic acid (produced during fermentation). Acidity is important in wine: as well as the flavor, it affects the color, and also the effectiveness of **sulfur dioxide** additions. It is commonly measured as total acidity, which is the sum of fixed and volatile acids. Acidity is also measured as **pH**, although this does not correlate exactly with total acidity. The relationship between the acid composition of wine and its actual acidic taste is a complex one: some acids are naturally more acidic than others for chemical reasons. The perception of acidity is also strongly influenced by other flavor components of the wine, notably sweetness, which counters the perception of acidity quite markedly.

Alcohol Common name for **ethanol**.

Aldehydes Also known, along with **ketones**, as carbonylated compounds. Rapidly combines with **sulfur**

dioxide in wine. These are formed whenever wine is exposed to oxygen. The most important aldehyde in wine is **acetaldehyde**. Other aldehydes present in wine can be important in terms of flavor development: some higher aldehydes contribute to wine aroma, and vanillin is a complex aromatic aldehyde that can be present in wine because of fermentation or aging in oak barrels.

Alleles A number of different forms of the same gene.

Amino acids The building blocks of proteins, present in wine at appreciable levels, and responsible for the taste of **umami**. There are just 20 of them, responsible for the many thousands of different proteins produced by living creatures.

Ampelography The science of vine identification.

Anthocyanins **Phenolic** compounds responsible for the color of red and black grapes. In wine, they interact with other components to form pigmented polymers and are responsible for wine color.

Antioxidant Chemical compound that prevents oxidation by reacting with oxidation: it takes the hit to protect other compounds.

Apiculate yeast Group of yeasts involved in wild or indigenous yeast fermentations.

Ascorbic acid Also known as vitamin C, and sometimes used in wine as an **antioxidant**. Acts synergistically with **sulfur dioxide**, but controversially the two together have been implicated in some incidences of **random oxidation**.

Astringency Perceived in the mouth by the sense of touch, astringency in wine is contributed by **tannins**—a drying, mouth-puckering sensation.

Autolysis In wine, the self-destruction of yeast cells, releasing flavor components into the wine.

Balling Another term for **Brix**.

Bâttonage Stirring the lees, the yeast-cell deposit at the bottom of a fermentation vessel.

Baumé A measure of dissolved compounds in grape juice, used as an approximate measure of sugar levels. Common in Europe. A degree of Brix is equal to around 1.75 percent sugar.

Bentonite A clay used to fine (remove suspended solids from) wine.

Biodynamic Controversial souped-up form of organic viticulture with a cosmic slant.

Bitterness Taste sensation, not all that common in wine, and commonly confused with **astringency** and sourness.

Botrytis Genus of fungus, but also a common term to describe infection of grapes by *Botrytis cinarea*. If it affects grapes that are already ripe, it can be beneficial and is responsible for many of the world's great sweet wines. But it also has a malevolent influence, causing gray rot.

Brettanomyces Yeast genus that is, at sufficient concentrations, a spoilage organism in wine. Controversial.

Brix A density scale of measuring sugar content, also known as **Balling** (the two are used interchangeably). Each degree represents one percent of sugar In the grape juice. As an example, 20 degrees Brix corresponds to about 12% alcohol in a finished wine (this will depend on the conversion of sugar to alcohol by the yeast).

Cane A one-year-old stem of a grapevine, used as the basis of cane (aka **rod and spur**) pruning.

Carbon dioxide The well-known gas, vital to plant growth as the carbon source of photosynthesis, and also contributing to global warming. Used in winemaking to protect grapes, must, and wine from oxygen. Naturally produced in fermentation, and while in most wines this dissipates, in some styles appreciable levels remain where it helps to preserve freshness.

Catechin A Flavan-3-ol (the other significant one in wine is epi-catechin), a group of **phenolic compounds** that are the building blocks of **tannins**. In their polymeric forms, where they are called procyanidins (often referred to as condensed tannins).

Chaptalization The addition of sugar to must to boost alcoholic strength. Allowed in some northern European wine countries.

Chloranisoles Group of compounds containing chlorine. Largely responsible for musty taint caused by rogue corks. Best-known is **2,4,6-trichloroanisole** (TCA).

Clonal selection Taking cuttings from a superior vine in the vineyard for further propagation. While vineyards are often planted by using genetically identical material (a single **clone** of a variety), over time, some vines perform better than others, beyond site-specific influences. This might be due to of spontaneous mutations but is commonly because of disease pressure. The most vigorous, actively growing vines in a vineyard are usually not the ones producing the best-quality fruit and should not automatically be chosen to take cuttings from.

Clone In viticulture, a group of vines all derived from the same parent plant by vegetative propagation (cuttings), and thus genetically identical.

Colloids Very tiny particles, smaller than a micrometer in diameter, usually made up of large organic molecules, that are important for the body of a wine. These can be removed by filtration, which can have the effect of stripping flavor from a wine if done clumsily.

Congeners Imprecise term for impurities in a spirit, thought to be responsible in part for hangovers.

Co-pigmentation Trendy but controversial term to describe the fixing of color in red wines by the presence of noncolored phenolic compounds. For example, one of the reasons Shiraz is sometimes fermented with a dash of Viognier is because co-pigment phenolics from the white grapes facilitate production of a darker wine.

Copper Element often present in the vineyard due to the use of Bordeaux mixture to combat fungal disease (which contains copper sulfate). Also used to remove volatile sulfur compounds from wine to prevent reductive taints. It is toxic to soil organisms, though.

Cordon Name for the woody arm or branch of a vine, growing horizontally from the main trunk and which bears spurs (when spur pruning is adopted).

Cover crop Plants grown between the vine rows during the dormant season. They are then plowed into the soil before the vine growth kicks in. Can be used to improve soil structure as well as adding nutrients and helps prevent erosion. But the cover crop can also compete with the vine for water.

Cross-flow filtration Also known as tangential filtration, this is the technique behind reverse osmosis. Increasingly adopted in the wine world, as it has many advantages over plate filters.

Cryoextraction The controversial use of freezing fresh grapes before crushing to extract only the sweetest, richest juice. Used by producers of sweet wines to enhance them, but some consider this to be cheating.

Cytokinins A group of **plant hormones**. Particularly involved in regulating cell division, thus affecting growth stages.

Dekkera The spore-forming form of the yeast Brettanomyces.

Diacetyl Common name for buta-2,3-dione, this is a ketone produced during fermentation or by the action of lactic acid bacteria. Smells buttery and slightly sweet.

Downy mildew Significant disease of vines caused by *Plasmopara viticola*. Introduced to Europe from the USA in the 1880s this caused significant damage, until it was managed by spraying Bordeaux mixture. Still a major problem.

Enzymes Proteins that catalyze chemical reactions (accelerate them, or reduce the temperatures needed for them to occur). Commercial preparations exist that can be used in winemaking for various reasons, some more justified than others.

Epigenetics A way in which the genome of an organism can be modified other than by changes in the coding sequence of the DNA. These changes can alter the way that genes are expressed (are turned into proteins) and are very important.

Esters Important to wine flavor. They are formed by the reaction of organic acids with alcohols and are formed during both fermentation and aging. **Ethyl acetate** (also known as ethyl ethanoate) is the most common ester in wine, formed by the combination of acetic acid and **ethanol**. Most esters have a distinctly fruity aroma. Some also possess oily, herbaceous, buttery, and nutty nuances.

Ethanal Another term for **acetaldehyde**.

Ethanol Common name for **ethyl alcohol**, even more commonly referred to as just alcohol.

Ethyl acetate A common **ester** in wine formed by the combination of **acetic acid** and **ethanol**.

Ethyl alcohol In volume terms, ethyl alcohol is the most important component of wine, and is produced by fermentation of sugars by yeasts. Alone, it does not taste of much, but the concentration of alcohol in the final wine has a marked effect on the wine's sensory qualities.

4-ethylguaiacol See **volatile phenols**.

Extraction In winemaking, the removal of **phenolic compounds** from the grape skin during the winemaking process.

Fan leaf virus Among the most common vine diseases. Caused by a range of viruses, it is a serious problem, and a good reason for planting with specially treated virus-free rootstock.

Fixed acids A term used to describe the nonvolatile acids **tartaric** and **malic**. Some acids are intermediate between volatile and fixed, though.

Flavonoids A large group of plant **phenolic compounds**, including pigments such as anthocyanins.

Flor A film of yeast cells that can form on the surface of a wine, important in the production of some styles of Sherry and Vin Jaune.

Fructose One of the main sugars in grapes, along with **glucose**.

GC-MS Stands for gas chromatography–mass spectrometry. A sensitive analytical technique for separating and identifying volatile compounds or gases.

Gibberellins A group of **plant hormones** important for influencing growth and development of vines. Sometimes applied artificially. Important in shoot elongation and release from dormancy.

GIS Stands for geographical information systems, in viticulture used to gather data about physical characteristics of a vineyard for **precision viticulture**.

Glucose Produced by photosynthesis, this is the most important sugar of the grape.

Glycerol Produced during fermentation, this is a polyol that can make a wine taste slightly sweeter, but which, contrary to popular opinion, does not affect viscosity.

Higher alcohols Also known as fusel oils, these are products of fermentation. Can contribute some aromatic character wines.

Histamine Chemical involved in allergic reactions in humans. Present in some wines, but not thought to be responsible for adverse reactions, sometimes termed wine "allergies."

Homoclimes Trendy New World viticultural term for areas with similar climates.

Hybrid vines Vines produced by crossing two different species. Also called interspecific crosses. Most usually, disease-resistant American species are crossed with *Vitis vinifera* vines to produce resistant vines without the typical foxy flavors of the American species. These are also often called American hybrids or French hybrids.

Hydrogen sulfide Smells like rotten eggs, and a potential spoilage element formed during fermentation. It is often caused by a nitrogen deficiency in the must.

Integrated pest management Known simply as IPM, this is an agricultural system that aims to reduce inputs of herbicides, fungicides, and pesticides via intelligent use.

Internode The part of a stem between two nodes (where the buds occur).

Ketones Usually produced during fermentation. β-damascenone and α- and β-ionone are called complex ketones and are thought to exist in grapes. β-damascenone smells roselike. α- and β-ionone occur most notably in Riesling grapes.

Laccase A polyphenol oxidase enzyme responsible for chemical oxidation of the grape must when it is present.

Lactic acid A softer-tasting acid produced by the bacterial metabolism of the harsher **malic acid** during **malolactic fermentation**.

Lactones Compounds that can be present in grapes, but which more often come from oak (the oak lactones). Sotolon is a lactone characteristic of botrytized wines.

Leafroll virus A problematic viral disease that can only be eradicated using virus-free planting material. It is frequently spread by an insect vector, the vine mealybug.

Maceration Important in red wine, this is the process of soaking the grape skins to remove **phenolic compounds**. There are various ways of doing this, including modern innovations, such as an extended maceration period at cool temperatures before alcoholic fermentation is allowed to start.

Maderization The process of a wine becoming oxidized—to become maderized, usually by heat.

Malic acid Along with **tartaric acid**, this is one of the two main organic acids in grapes. Transformed to the softer **lactic acid** by the action of lactic acid bacteria during **malolactic fermentation**.

Malolactic fermentation The conversion of **malic acid** to **lactic acid** effected by lactic acid bacteria.

Mercaptans Volatile sulfur compounds sometimes found in wine. Also known as thiols. Can be bad, but some of them are good. It depends.

Methoxypyrazines Nitrogen containing heterocyclic compounds produced in grapes by the metabolism of **amino acids**. 2-methoxy-3-isobutylpyrazine is the compound responsible for the bell pepper, grassy character common in Sauvignon Blanc and Cabernet Sauvignon wines. Concentrations of methoxypyrazine decrease with ripening; sunlight on grapes also reduces the concentration.

Microoxygenation Technique of slow oxygen addition to fermenting or maturing wine.

Monoterpenes Chemical group adding to the aroma and flavor of varieties such as Muscat and also Riesling.

Mouthfeel In vogue tasting term used to describe textural characters, most specifically structure, of a wine.

Nodes The parts of the grapevine stem that contain the bud structures, separated by internodes.

Oeschle Measure of sugar concentration in grapes.

Oenological tannins Commercially produced tannins, normally not of grape origin, used to add to wines. More commonly practiced than you would think.

Oidium Fungal disease commonly known as **powdery mildew**. Caused by *Erysiphe necator* (also known as *Uncinula necator*).

Oxidation Substances are oxidized when they incorporate oxygen and lose electrons or hydrogen. Oxidation is always accompanied by the opposite reaction, **reduction**, such that when one compound is oxidized, another is reduced. In wine, oxidation occurs on exposure to air and is almost always deleterious.

Pectins Carbohydrates that cause the walls of plant cells to adhere to each other.

pH Technically, the negative logarithm of the hydrogen ion concentration in a solution. A scale used to measure how acid or alkaline a solution is (more acid = lower pH). Very important in winemaking.

Phenolic compounds Also known as **polyphenols**, this is a large group of reactive **polymers** with the phenol group as the basic building block. Important in wine.

Physiological ripeness Also known as phenolic ripeness. Fashionable term used to distinguish the stage of maturity of the vine and grapes, as opposed to just the sugar levels. In warm climates, picking by sugar levels alone can result in unripe, herbaceous traits in the wine.

Phytoalexins Important antimicrobial compounds produced by plants in response to attack.

Pigmented polymers Also known as pigmented tannins. Complex group of chemicals now implicated in red wine color production. Formed when **catechins** and **anthocyanins** combine during fermentation.

Plant hormones A group of signaling molecules that influence plant growth and development, and also send out stress responses. Known in the trade as plant growth regulators. This group includes auxins, **cytokinins**, **gibberellins**, **abscisic acid**, and **ethylene**. Some researchers also include the brassinosteroids and jasmonic acid in this club.

Polysaccharides Carbohydrate **polymers** with sugars (monosaccharides) as the main subunits.

Polymers Molecules that form as the result of **polymerization** of smaller subunits.

Polymerization Making larger molecules by joining together smaller subunits.

Polyphenols These are probably the most important flavor chemicals in red wines but are of much less importance in whites. Polyphenols are a large group of compounds that use phenol as the basic building block. An important property of **phenolic compounds** is that they associate spontaneously with a wide range of compounds, such as **proteins** and other phenolics, by means of a range of noncovalent forces (for example, hydrogen bonding and hydrophobic effects).

Polyphenol oxidase (PPO) An **enzyme** that causes browning by reacting with **phenolic compounds** and oxidizing them. Grapes that have been infected with fungi have high levels of PPO and this is can cause **oxidation** of the wine, both directly and by combining with free **sulfur dioxide**.

Powdery mildew Also known as oidium. Nasty fungal disease that devastated European vineyards when it was introduced from the USA in the late 1840s. Was eventually countered by dusting vines with sulfur, a treatment that is still used today by some. The fungus responsible is *Uncinula necator*.

Precision viticulture Selectively applying vineyard inputs according to relevant data.

Proteins Polymeric molecules made from combinations of the 20 naturally occurring **amino acids**. Proteins are encoded by genes.

PVPP Shorthand for poly(*N*-vinylpyrrolidinone), sometimes used to remove bitter **phenolic compounds** from white wine.

Quercitin A common **flavonol** in grapes.

Quercus The genus *Quercus* comprises the various species of oak.

Random oxidation Also known as "sporadic post-bottling oxidation," describes the premature browning that occurs to some white wines some months after bottling. It is a problem common enough that some industry figures have referred to this as the "new cork taint." The main explanation for this phenomenon is variable oxygen transfer through the cork. Wines are protected against **oxidation** through the addition of **sulfur dioxide** at bottling. If the free sulfur dioxide falls to very low levels then the wine is unprotected, and browning can occur. Corks have been shown to vary dramatically in their oxygen-transfer properties, and random oxidation is thought to be caused by the subset of corks that let in significantly more oxygen than others. However, some scientists suspect that random oxidation may be caused by as yet poorly understood chemical reactions independent of the closure. It has been suggested that common winemaking addition of the **antioxidant ascorbic acid** to keep white wines fresh may have the paradoxical effect of rendering the added sulfur dioxide less effective and making some wines susceptible to oxidation. Another proposed cause of random oxidation is poor procedure or intermittent failure on the bottling line, allowing some wines to have much higher levels of dissolved oxygen from the outset. Random oxidation is mainly a problem with white wines: while oxygen ingress through the closure will certainly damage red wines, they are more resistant to oxidation because of their high **phenolic** content. Oxidation is also more likely to be spotted in white wines because of the dramatic color change that accompanies it.

Redox potential Stands for **reduction–oxidation** potential. Can be measured but is not considered a terribly useful measurement in winemaking because it can change very fast.

Reduction Shorthand term used to refer to sulfur compound flavors in wine. Technically, it is the opposite of **oxidation**.

Resveratrol Phenolic compound that is also a **phytoalexin**, present in grapes and red wines. May have some health-enhancing properties.

Reverse osmosis Controversial filtration technique used to concentrate wines and also for alcohol removal.

Rod and spur Alternative name for **cane** pruning.

Saccharomyces cerevisiae The yeast species responsible for alcoholic fermentation, known colloquially as brewer's or baker's yeast. It appears as many different strains.

Saignée Also known as vat bleeding, a winemaking technique for taking juice of skins to increase the solids to juice ratio in red winemaking, thus enhancing the **phenolic** content of the resulting wine.

SO₂ Chemical formula for **sulfur dioxide**.

Sorbitol An alcohol present in low levels in wine with a hint of sweetness to it, occasionally used by fraudsters as an illegal addition.

Sotolon A **lactone** present in botrytized wines.

Spinning cone Technically a gas–liquid counter current device. A way of stripping alcohol from wines without removing important volatiles, used commonly to concentrate fruit juices without taking out interesting bits. Widely used, but because of the expense of the equipment this is just a service industry.

Spur A stubby grapevine shoot pruned back to just a few nodes.

Spur pruning A way of pruning vines that leaves just short **spurs** on a permanent **cordon**.

Succinic acid An acid present at low concentrations in grapes and wine.

Sulfides Reduced sulfur compounds that occur during winemaking. Usually negative but can be complexing at the right levels.

Sulfur dioxide (SO₂) Hugely important molecule added in winemaking to protect wine from oxygen and microbes.

Tangential filtration See cross-flow filtration.

Tannins Tannins are found principally in the bark, leaves, and immature fruit of a wide range of plants. They form complexes with **proteins** and other plant **polymers** such as **polysaccharides**. It is thought that the role of tannins in nature is one of plant defense. Chemically, tannins are large polymeric molecules made up of linked subunits. The monomers here are **phenolic compounds** that are joined together in a bewildering array of combinations and can be further modified chemically in a myriad of different permutations. There are two major classes of tannins: condensed and hydrolyzable. Hydrolyzable tannins are not as important in wine. The condensed tannins, also known as proanthocyanidins, are the main grape-derived tannins. They are formed by the **polymerization** of the polyphenolic flavan-3-ol monomers **catechin** and epicatechin.

Tartaric acid The most important grape-derived acid in wine. Often precipitates out as harmless tartarate crystals.

TCA Abbreviation for 2,4,6-trichloroanisole.

Terpenes A large family of compounds widespread in plants. Grapes contain varying amounts, which survive vinification to contribute to wine odor. More than 40 have been identified in grapes, but only half a dozen are thought to contribute to wine aroma. They are highest in Muscat wines: the distinctive floral, grapey character is down to the likes of linalool and geraniol. Other varieties such as Gewürztraminer and Pinot Gris also have a terpene component to their aromas.

Total acidity An important measurement for winemakers, made by titration. Given in terms of g/l of **tartaric** or sulfuric acid. It includes measurement of both fixed and volatile acids.

2,4,6-Trichloroanisole Potent, musty-smelling compound largely responsible for cork taint.

Umami The fifth basic taste, only recently recognized. It is the taste of "savoriness" and results from the detection of **amino acids**.

Veraison The process in grape ripening where the skins begin to soften and the berries change color (most obvious in dark-skinned grapes). It is the transition from berry growth to berry ripening. Post-*veraison*, acidity decreases and sugar begins to accumulate.

Vitis vinifera The species name for the Eurasian grapevine, to which the varieties we know and cherish all belong.

Volatile acidity The acidity contributed by the various volatile acids, the most significant of which is **acetic acid**. A little is okay, but too much is very bad and makes the wine smell of vinegar.

Volatile phenols Important in wine aroma. 4-ethylphenol and 4-ethylguaiacol, found predominantly in red wines, are formed by the action of the spoilage yeast Brettanomyces and have distinctive gamey, spicy, animally aromas; 4-vinylphenol and 4-vinylguaiacol are rare in red wines and more common in whites, and also have largely negative aromatic properties.

Yeasts Unicellular fungi important for fermenting grape juice to wine. The unsung stars of the wine world.

Index